P9-AQT-418

WILD
CARIBBEAN

WILD CARIBBEAN

THE HIDDEN WONDERS OF THE WORLD'S MOST FAMOUS ISLANDS

MICHAEL BRIGHT

YALE UNIVERSITY PRESS NEW HAVEN & LONDON

Gulf of Mexico
U.S.A
Grand Bahama
Little Abaco
Northwest Providence Channel
Great Abaco
Bimini Islands
Berry Islands
Northeast Providence Channel
Eleuthera Island
New Providence
NASSAU
BAHAMAS
Cat Island
San Salvador
Straits of Florida
Santaren Channel
Andros Island
Cay Sal
Great Guana Cay
Conception Island
Rum Cay
Great Exuma Island
Long Island
Samana Cay
LA HABANA (HAVANA)
Guanabacoa
Canal Nicholás
Matanzas
Canal de Yucatan
Pinar del Río
Sagua la Grande
Canal Viejo de Bahama
Jumentos Cays
Crooked Island
Santa Clara
Placetas
Cienfuegos
Sancti Spíritus
Morón
Duncan Town
Acklins Island
Merida
Cancun
Isla de la Juventud
Trinidad
Ciego de Avila
CUBA
Archipiélago de los Canarreos
Florida
Camagüey
YUCATAN PENINSULA
Isla Cozumel
Archipiélago de Jardines de la Reina
Las Tunas
Banes
Holguín
Golfo de Guacanayabo
Manzanillo
CAYMAN ISLANDS (to UK)
MEXICO
Sierra Maestra
Guantánamo
Little Cayman
Cayman Brac
Pico Real del Turquino 1974m
Santiago de Cuba
Grand Cayman
Windward
Banco Chinchorro
7680m
GUANTANAMO BAY (to US)
NAVASSA ISLAND (to US)
Ambergris Clay
Montego Bay
BELIZE CITY
Blue Mountain Peak 2256m
JAMAICA
Turneffe Is.
Spanish Town
KINGSTON
Is. Santanilla (to Hon.)
BELIZE
GUATEMALA
Is. de la Bahia
Golfo de Honduras
Pedro Cays (to Jam.)
Punta Patuca
Bajo Nuevo (to Col.)
Laguna Caratasca
Mosquito Coast
HONDURAS
Carib
TEGUCIGALPA
Cayos Miskitos (to Nic.)
EL SALVADOR
I. de Providencia (to Col.)
NICARAGUA
Punta de Perlas
MANAGUA
I. de San Andrés (to Col.)
Isla del Maiz (to Nic.)
Lago de Nicaragua
Santa Marta
Barranquilla
COSTA RICA
SAN JOSÉ
Limón
Cartagena
Archipelago de San Blas
Bocas del Toro
Laguna de Chiriquí
PACIFIC OCEAN
Colon
PANAMA CANAL
Serrania del Darien
Golfo del Darién
PANAMA
PANAMA
COLOMBIA
Golfo de Panama

0 100 200 kilometres
0 100 200 miles
ATLANTIC OCEAN
TURKS & CAICOS ISLANDS (to UK)
Mayaguana
Caicos Passage
Caicos Islands
Grand Turk Island
Little Inagua
Turks Islands
Great Inagua
Hispaniola
Île de la Tortue
Cap-Haïtien
Gonaïves
HAITI
Pico Duarte 3175m
Santiago de los Cabelleros
San Francisco de Macorís
DOMINICAN REPUBLIC
La Romana
Pic la Salle 2680m
Barahona
SANTO DOMINGO
Isla Saona
Isla Beata
Isla Mona
Canal de la Mona
Mayagüez
Puerto Rico Trench
PUERTO RICO (to US)
SAN JUAN
Bayamón
Caguas
Ponce
St Thomas
VIRGIN ISLANDS (to US)
St Croix
BRITISH VIRGIN ISLANDS (to UK)
Anegada
Tortola
Virgin Gorda
Anegada Passage
ANGUILLA (to UK)
Anguilla
St-Martin (to France)
Sint Maarten (to Neth.)
St-Barthélémy (to Fr.)
Saba
St Eustatius
Barbuda
ANTIGUA & BARBUDA
ST JOHN'S
Antigua
St Kitts
Nevis
BASSETERRE
ST KITTS & NEVIS
NETHERLANDS ANTILLES (to Neth.)
MONTSERRAT (to UK)
Leeward Islands
La Soufrière 1467m
GUADELOUPE (to France)
BASSE-TERRE
Grand-Bourg
I. des Saintes
Dominica Passage
DOMINICA
ROSEAU
I. de Aves (to Ven)
Lesser Antilles
Martinique Passage
MARTINIQUE (to France)
Mt. Pelée 1397m
FORT-DE-FRANCE
St Lucia Channel
CASTRIES
ST LUCIA
BARBADOS
BRIDGETOWN
Saint Vincent Passage
La Soufriere 1234m
St Vincent
KINGSTOWN
ST VINCENT & THE GRENADINES
The Grenadines
Windward Islands
GRENADA
ST GEORGE'S
bean Sea
Lesser Antilles
ARUBA (to Netherlands)
NETHERLANDS ANTILLES (to Netherlands)
WILLEMSTAD
Curaçao
Bonaire
Los Roques (to Ven.)
Las Aves (to Ven.)
Orchila (to Ven.)
Blanquilla (to Ven.)
Los Hermanos (to Ven.)
Los Testigos (to Ven.)
Isla de Margarita (to Ven.)
Isla La Tortuga (to Ven.)
Dragon's Mouth
Tobago
TRINIDAD & TOBAGO
PORT-OF-SPAIN
Trinidad
San Fernando
Serpent's Mouth
Ríohacha
Golfo de Venezuela
CARACAS
Sierra Nevada de Santa Marta 5800m
Orinoco Delta
VENEZUELA

FINDING

PARADISE

ANY MENTION of the Caribbean is sure to conjure up stunning images of the world's finest beaches, fringed with swaying palms and lapped by an azure sea – a tropical paradise. As a holiday destination, there are few places to match it, but not far from the bars and the recliners is another Caribbean – the wild one. Here, birds with gaudy feathers and butterflies with see-through wings live in lush, tropical forests filled with orchids and bromeliads; multicoloured coral reefs bustle with vibrant fishes and menacing reef sharks; white, pink and black sand beaches play host to nesting sea turtles; underground labyrinths of river-filled caverns and seemingly bottomless sinkholes are the secret hideaways for bats, bugs and blind cavefish; prehistoric iguanas and countless lizards scramble among exotic cactuses and thorny scrub; salt ponds and freshwater lakes are graced by coral-coloured flamingos; and the tangled lattice of mangroves is not only home to tree-climbing crabs and roosting seabirds, but also a safe nursery for myriad tiny marine creatures.

Silver waterfalls cascade over dark, rocky ledges and sheer white cliffs; at night the water in phosphorescent bays glows an eerie blue-green; blowholes spout fountains of spray high into the air and blue holes lead the experienced diver to hidden worlds long forgotten or yet to be discovered. Mysterious lakes boil, and sulphur springs bubble and belch, a gentle reminder of the powerful forces that created many of these islands; and there are the volcanoes themselves – glowing, smoking, shaking, some actively and violently erupting.

There are places with intriguing names – Bimini Wall, Boiling Lake, Cockpit Country and Virgin Gorda Baths – all waiting for the inquisitive traveller. This is a place not to sit but to explore, and in this brief exploration of the Caribbean we offer a taste of what is in store; after all, it was the unspoiled wilderness, not a tourist mecca, that was first encountered by the early European voyagers.

When the explorer and trader Christopher Columbus stepped ashore at what he named San Salvador on 12 October 1492, he must indeed have thought he was in paradise. The natives, who called the place Guanahani, were friendly; the flat, semi-arid islands, dominated by evergreen bush and thicket and surrounded by jagged coral reefs and shallow saltwater lagoons, were an undeveloped wilderness. The current San Salvador, however, was not necessarily his first landing place, for the precise island on which Columbus first set foot is unclear. Several vie for that honour, the two current favourites being Samana Cay, a small island in the central Bahamas, and the western island of Plana Cays, located to the east of Acklins Island and to the west of Mayaguana.

Silver waterfalls cascade over dark, rocky ledges and sheer white cliffs

Although Italian by birth, Columbus sailed under the flag of Castille. His three ships the *Niña*, the *Pinta* and the *Santa Maria* set sail from Palos de la Frontera in southwest Spain on 3 August 1492. The journey across the Atlantic was relatively uneventful, apart

PREVIOUS PAGES White sand beach and waving coconut palms at Bottom Bay in St Philip, Barbados.

OPPOSITE Tobago Falls drop 54 metres (177 feet) in four cascades into a deep, green pool and are a good place to see kingfishers *Chloroceryle americana croteta*.

from the crew becoming petulant after being becalmed and spending many weeks at sea without any sign of land. According to the ship's log, the first hint that Columbus was anywhere near land came on 7 October. Throughout the day, an immense flock of birds passed overhead, coming from the north and heading southwest. The crew ensnared a few, and it was plain that they were not seabirds and therefore could not rest on the sea.

OPPOSITE An aerial view of just a few of the fifteen dry, subtropical, raised islands in the British Virgin Islands group.

BELOW The red-lored parrot *Amazona autumnalis* is native to the humid forests of tropical Central and South America.

It was the expedition's sixty-fifth day in the Atlantic, so, believing the flocks were about to make landfall, the trio of ships followed. Today's ornithologists, comparing migration patterns with the dates in Columbus's journal, suggest that the birds were probably Eskimo curlews *Numenius borealis* and American golden plovers *Pluvialis dominica* on their autumn (fall) migration from North to South America. At that time, millions of these curlews – or 'prairie pigeons', as they were known on account of their extraordinary numbers, reminiscent of passenger pigeons – made the journey from Canada to Argentina. The great flocks stopped off in the Caribbean to rest, but nowadays, if recent sightings are to be believed, there is just a handful of Eskimo curlews left. The species might even be extinct, but in 1492 they were probably the most abundant birds to visit the archipelago.

What Columbus had observed, however, was more than a wildlife spectacle. He had witnessed the importance of the Caribbean as both a stopover and refuelling site for passage migrants and an overwintering site for seasonal residents. While the Eskimo curlew may not be able to excite the modern birdwatcher as it did Columbus and his crews, comparable natural events can be seen today at migration-watch sites throughout the region. In southeast Costa Rica, for example, the Kekoldi Indigenous Reserve near Puerto Viejo de Limón is overflown each year by up to 3 million birds of prey – vultures, kites, ospreys, hawks and falcons. At peak migration times, 200,000 birds may be counted in a single day, making this one of the most concentrated raptor sites in the world.

The seasonal bird traffic through the Caribbean is immense. In all, about 160 species of neotropical migrants breed in North America during the summer, but spend their winters in Mexico, Central America, South America and the Caribbean; the islands are host to about 120 of them. This means that 20 per cent of all the known bird species in the region, including many threatened species such as the Cape May warbler *Dendroica tigrina*, northern parula *Parula americana* and Bicknell's thrush *Catharus bicknelli*, are North American emigrants

What is surprising, however, is not the number of visitors, but the remarkable percentage of resident plants and animals that are unique to the islands. Of the 560 recognized bird species, 148 are endemic, 105 unique to single islands. The predominantly green-feathered, red-necked Amazon parrot *Amazona arausiaca*, for example, is confined to Dominica, while its close relative, the similarly green St Lucia Amazon *A. versicolor*, is found only on St Lucia.

ABOVE The diminutive northern parula warbler is one of many song birds that spend their winters in the Caribbean, but migrate to North America in the summer.

The same is true for reptiles and amphibians. The Jamaican snoring frog *Osteopilus crucialis*, as its common name suggests, comes only from Jamaica, as does the Jamaican laughing frog *O. brunneus*.

In fact, the Caribbean is a hotbed of endemism. Of the 189 known species of amphibians on the islands, 164 are endemic – an incredible 87 per cent. Some 418 of the 497 species of reptiles recorded (84 per cent) are endemic, as are 49 of the 164 mammals, many of them bats or rodents. In the plant world, the figure is about 6650 out of 12,000 recognized species.

Endemism, however, is a double-edged sword. While each creature is well adapted to live on its particular island, this does make it vulnerable to environmental change and, with the 35 million-strong human population on the Caribbean islands growing at more than 2.5 per cent each year, change is inevitable. About 48 species of endemic birds, 18 mammals and a staggering 143 amphibians are thought to be seriously endangered. Since the 1500s, during the centuries when wave after wave of European colonists followed Columbus, at least 38 species are known to have disappeared altogether.

In view of this terminal decline in the Caribbean's wildlife, Conservation International has highlighted the islands as one of the world's top six 'biodiversity hotspots', in the hope that some species may be saved; after all, the region harbours some extraordinary record breakers.

The world's smallest bird – the diminutive bee hummingbird *Mellisuga helenae* – lives on Cuba, an island that Columbus explored on his first voyage in 1492. But he could be excused for missing a creature no more than 5 cm (2 inches) long, and he probably missed the smallest frog (which we shall meet in chapter 1) too. The smallest snake is also found in the Caribbean, on islands such as St Lucia and Martinique. The longest-known Lesser Antillean threadsnake *Leptotyphlops bilineatus* is just 10.8 cm (4¼ inches) long and so thin that it could replace the lead in a pencil. And as more and more species are identified, world records are being broken thick and fast.

The isolated island of Beata, off the southwest coast of the Dominican Republic, is home to what is believed to be the world's smallest lizard. A mere 16 mm (just over ½ inch) long, the Jaragua sphaero *Sphaerodactylus ariasae* (a dwarf gecko) lives in moist leaf litter in a sinkhole and caves a couple of kilometres (just over a mile) from Punta Beata. It was discovered in 2001, and its discoverers are turning up new species in the region all the time. The herpetology teams from Pennsylvania State University and the University of Puerto Rico had previously named over 50 new species of amphibians and reptiles. In order to find tiny creatures like these, the trick apparently is to explore rugged places far from civilization. It just goes to show that, even in these crowded islands, there are many more natural secrets yet to be revealed to us.

Curiously, the lizards on each Caribbean island have evolved quite separately.

BELOW This tiny frog, one of the world's smallest at just 12 mm (½ inch) long, was spotted in Cuba's Alexander von Humboldt National Park.

At one time it was thought that any 'specialist' arose on one island and then migrated and found similar niches on others. However, recent DNA research at Washington University, studying the islands' anole lizards, reveals that this is probably not the case. The Luquillo Forest of Puerto Rico, for example, is home to several habitat specialists: there is an anole species with short legs that crawls unhurriedly on narrow twigs; another with long legs that runs rapidly across the ground; and one with large, clinging toe-pads that lives high in the trees. The same anole specialists are to be found on near neighbours Cuba and Jamaica, but on each of the islands they are of completely different species – a classic example of convergent evolution, when unrelated animals develop the same adaptation because they occupy similar niches. In this case, entire communities have converged.

How, though, did these animals make it to islands that are somewhat distant from any mainland? The answer is a long one – 150 million years long – and it began thousands of kilometres to the west.

The Caribbean Sea actually started out in the Pacific Ocean – that is, if one of the current theories about its formation is to be believed. The problem is that the Caribbean is a geologically complex region, and any one of many theories explaining its origins could be correct. The Pacific theory suggests that the Caribbean plate, one of the plates that slides about on the Earth's surface during the process known as continental drift, was created in the eastern Pacific about 130–150 million years ago, a time when brontosaurs were thundering about the land and plesiosaurs were terrorizing the sea. At this time, North and South America were two separate continents, with much of modern Central American under water.

An arc of volcanic islands, labelled the Proto-Antilles, preceded the Caribbean plate's leading edge, stretching from west of where Mexico is today southwards to Ecuador, but they did not remain there for long. By about 80 million years ago they had 'drifted' to the northeast, squeezing through the gap between North and South America and creating a string of islands that loosely connected the two continents.

Like many seafarers, the islands carried their cargo of stowaways. *Eleutherodactylus* frogs ventured north from South America, so that today more than 140 species of little olive-brown frogs are spread throughout the Caribbean. They improved their chances of establishing themselves on the islands by dispensing with the need to be near open water to reproduce. Instead, their tadpoles develop in a sphere of water or in foam nests and emerge as young, fully formed froglets. Today, they are *the* frogs of the Caribbean. Jamaica, Cuba and southern Hispaniola have small, ground-dwelling frogs with short, high-pitched calls, and northern Hispaniola and Puerto Rico have larger tree-dwellers with longer, low-

North and South America were two separate continents, with much of modern Central America under water

pitched calls. Some of their early ancestors have been found on Dominica, preserved in amber that is 25 million years old.

Ancient pine trees progressed south from North America, along with feeble-flying butterflies, such as the orange and black pierid butterfly *Dismorphia* spp., which is such a poor flyer that it can be found on the larger, ancient islands of Cuba, Hispaniola and Puerto Rico, but not on more recently formed islets a few hundred metres offshore. The forests were, and still are, home to glasswing butterflies *Cithaerias* spp. With their flimsy, transparent wings, they would never have been able to cross any expanse of sea to populate the islands, so they must have flopped aboard when the Proto-Antilles brushed North America during the drift to the east. Today, about 350 known species of butterfly are found only in the Caribbean, again with many endemics restricted to single islands.

ABOVE The Allen's Cay iguana *Cyclura cychlura inornata* is well known to tourists in the Bahamas because boat operators stop to visit the only natural breeding populations at Leaf Cay and U Cay.

As the plate of continental crust, with its island chain, progressed northeastwards, it separated from the mainland, but it was already carrying a fair number of passengers, and more were joining all the time. Following close behind the insects and frogs were more insect-eaters. An ancient tody, a small insect-eating bird, probably made the short hop across the sea from North America. It was the ancestor of a whole bunch of todies, each species today confined to its own island in the Greater Antilles.

The large, shrew-like insectivore *Solenodon* was already on board. This ancient mammal, which in all probability shared the mainland with the dinosaurs, was once common in North America, but after the demise of its giant neighbours it was ousted by the newly evolving modern mammals and survives today only on Cuba and Hispaniola.

In fact, one of the causes put forward to explain the disappearance of the dinosaurs and their relatives worldwide probably had a profound effect on the entire fauna of the embryonic Caribbean. It was almost wiped out before it even got going. There is good evidence to suggest that a large extraterrestrial body, perhaps a comet or asteroid, slammed into the Earth about 64 million years ago, and its precise point of impact is thought to have been the Yucatán Peninsula,

then submerged by a shallow sea. An enormous tsunami would have spread across the region, with cataclysmic consequences.

Nevertheless, animals living in the highlands did pull through, and they continued to travel on their island 'rafts'. Then, about 55 million years ago, things came to an abrupt halt. Cuba bumped into the Bahamas Bank, a submerged platform of carbonate rock in the western Atlantic Ocean, gaining more than half its size again in limestone. Today part of the island is limestone and part original volcanic rock. The other islands of what are now the Greater Antilles kept going. About 35 million years ago Puerto Rico broke away from Hispaniola, and some 10 million years later Hispaniola and Cuba, which had been a single large island, separated.

Freshwater fishes, which had joined the floating islands when they were squeezed between North and South America, failed to roam very far along the chain. Cuba retained the lion's share, with a great diversity including its unique cavefish *Lucifuga* spp. and the long and slender garfish *Atractosteus tristoechus*. Hispaniola managed to keep just one family of live-bearing fish – the Poeciliidae – that went forth and multiplied. Nowadays they fill every available ecological niche on the island – streams, ponds, salty lakes, sulphurous springs and mangroves. Puerto Rico lost out altogether. Most likely, the freshwater fishes failed to move far before the split with Hispaniola, and today the smaller island has none.

While the large islands of the Greater Antilles were assembling into their present positions, the rest of the group to the southeast disappeared below the waves, to be replaced by a new arc of ultra-violent volcanoes that were thrust towards the surface: new islands were born. Mountains with names like Pelée and Soufrière were to bring terror to paradise. These monster volcanoes now form the island arc of the Lesser Antilles along the eastern edge of the Caribbean plate, the oldest in the north and the youngest in the south, and they are still full of fizz, the Soufrière Hills volcano on Montserrat bursting into life in 1995 and still erupting today.

Some volcanoes never made it to the surface or were eroded by the pounding Atlantic surf. Coral reefs grew on their submerged crater rims, and these eventually formed islands of pure white limestone. In the northern part of the Lesser Antilles today they are lined up in two parallel chains, with the volcanic islands bordering the Caribbean having black sand beaches, and the limestone islands facing the Atlantic white.

These two rows of smaller islands were never close, let alone connected, to the mainland, so plants and animals had to reach here by other means. Rafting on floating vegetation was popular. The obvious passengers were insects, spiders and other invertebrates that floated to the islands from estuaries like the Orinoco, on the northeast coast of Venezuela, but there were also small mammals. The Caribbean islands once had more than 55 species of rodents whose ancestors must have rafted from the mainland. Many are now extinct, but their fossil remains are being found in caves all over the region.

OPPOSITE The bottlenose dolphin, seen here swimming over a coral reef in Belize, is one of several species of dolphins to be found in the Caribbean.

ABOVE The glorious view of Petit Piton, viewed above Soufriére, St Lucia.

People, of course, arrived on more sophisticated 'rafts', although if they were anything like Columbus they would have had no idea where they were. He believed he had sailed to the Indies, a term that in those day referred to the lands in southern and eastern Asia, which is why these islands eventually became known as the West Indies. Politically, they consist today of 12 independent nations and numerous islands regulated by Britain, France, the Netherlands and the USA. The largest are Cuba, Hispaniola, Jamaica and Puerto Rico, collectively known as the Greater Antilles, which together account for 90 per cent of the landmass; the rest is a collection of over 7000 smaller islands, islets, cays and reefs.

The name Caribbean came not from the Europeans but from the Caribs, one of the dominant local tribes at the time of the European invasion. When Columbus arrived, three main indigenous Amerindian groups inhabited the islands – the Arawak or Taíno on the Greater Antilles and the Bahamas, the Caribs on the Lesser Antilles and the Ciboney in Cuba.

The first people to colonize the islands had come much earlier. About 7000 years ago they arrived from South America and settled on islands in the southern Caribbean, such as Trinidad. Evidence for their occupation comes in the form of shell middens, fishing tackle made from bones, grinding stones and stone tools found in an archaic or pre-ceramic archaeological site at Banwari Trace in southwest Trinidad. From here, they migrated northwards through the Lesser Antilles, evidence having been unearthed at sites on St Kitts, Nevis, Antigua, the US Virgin Islands, Hispaniola and Cuba.

Since then various cultures have moved into the islands, and, by 800 CE, the Arawak had arrived. A gentle race, they lived in villages of 1000–5000 inhabitants controlled by chiefdoms; they grew maize, yucca and cassava, hunted and gathered shellfish, played in ball courts and celebrated on large plazas.

A hundred years before Columbus arrived, the Arawak's peace was shattered. In the Lesser Antilles, they were displaced by the Caribs, who were also from the Orinoco area. Unlike the Arawak, the Caribs were aggressive warriors, but they were also accomplished boat builders and expert navigators, and they were the source of the gold – which they did not smelt themselves but obtained by trading with the South American mainland – that the Spanish eventually cherished. It is said that only the men spoke Carib, the women speaking Arawak. Legend has it that Carib raiding parties took female Arawak prisoners as their wives. They were also said to be cannibals, with the Carib word *karibna,* meaning 'person', corrupted by the Spanish to *caníbal*. The assumption that cannibalism was rife among the Caribs, however, must be treated with a certain reservation, as Queen Isabella of Castile, Columbus's patron, had ruled that the Spanish could take only slaves who were convicted cannibals. Many tribes, whether cannibals or not, were therefore conveniently labelled as such.

Having Europeans on the doorstep had a devastating impact on all the native peoples. When Columbus arrived, the Taíno population is thought to have numbered between 100,000 and 400,000, but by the end of the eighteenth century there were fewer than a few hundred surviving. They were taken into slavery – Columbus wrote in his log, 'with fifty men we could subjugate them all and make them do whatever we want'; they were worked and overworked on plantations, farms and in the mines; and they were murdered in countless numbers. They also fell victim to diseases, such as smallpox, to which they had no resistance.

At first the Spanish slave-masters focused on the Ciboney of Cuba, but after these died out they turned on the Taíno, and when the latter fell into short supply slaves were sought in Africa. It was the start of the Slave Triangle, in which the Spaniards were just the tip of an iceberg. Traders set out from European ports and headed for the west coast of Africa, where they traded goods for slaves. The slaves were packed tightly into ships and transported for six to eight weeks, under dreadful conditions, to the Caribbean. It is thought that 8 million slaves died to bring 4 million to the region, the death rates on Bristol ships, according to the Reverend Thomas Clarkson, who in 1787 examined the records of the Merchant Venturers, being higher than most. Nevertheless, the traders of Liverpool, London and Bristol, with their Spanish, Portuguese, Dutch, Danish and French counterparts, became rich.

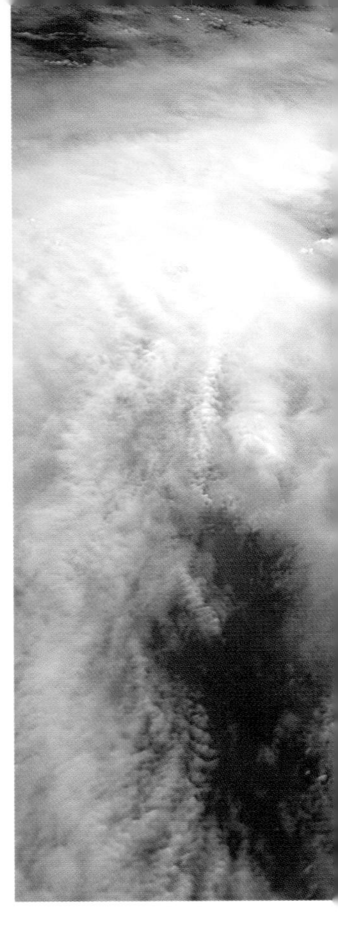

Most of the slaves worked on the land. Their European masters had it cleared of its natural vegetation, including magnificent stands of mahogany trees that were chopped down and transported to Europe for ship building, cabinet making and eventually to make railway carriages. In their stead they created huge plantations, growing sugar cane, coffee, rice and tobacco, all of which were exported to Europe, along with other goods, such as rum, a by-product of the sugar-making process. Many islands were laid bare, and even on the others very little virgin forest survives intact today.

That anything survives at all is something of a miracle, for between July and November each year the entire region is on 'hurricane alert', braced for the most powerful storms on Earth. Exceptionally high winds, extraordinary rainfall and battering storm surges, charged with as much energy as 10,000 nuclear bombs, hit the Caribbean with unfailing regularity.

The word hurricane, in fact, has its origins here: the name of the Central American god of evil, Hurucán, means 'storm' in the Arawak language, and today the term refers to the tropical storms that occur in the North Atlantic and eastern Pacific.

But hurricanes do not start in the Caribbean. They develop on the other side of the Atlantic, off the coast of West Africa, where conditions are ripe for the formation of high-intensity tropical storms – high humidity, light winds and warm sea-surface temperatures. The hurricanes travel westwards across the ocean, accompanied by winds that gust to 330 km/h (200 mph) and a storm surge that can see the sea level rise up to 10 metres (33 feet), producing a dome of water perhaps 160 km (100 miles) across that can swamp the land.

The deadliest-known hurricane hit the region in October 1780, at the time of the American Revolution. In its path were the islands of Martinique, St Eustatius and Barbados, and sheltering in their harbours and bays were British and French naval fleets fighting for control of the region. This storm caused far more human fatalities than any storm ever documented. On the islands themselves an estimated 22,000 lives were lost, and many thousands more at sea.

In more recent years, many monster hurricanes have hit the islands. In September 1930 one swept across Hispaniola, destroying the city of Santo Domingo in the Dominican Republic. In 1963 Hurricane Flora hit Haiti and Cuba, and in September 1974 Hurricane Fifi slammed into Honduras, Belize, Guatemala and Mexico before heading towards the Pacific. In 2004 Hurricane Ivan blasted Grenada, causing catastrophic damage, then turned towards Jamaica and Grand Cayman. Things are getting worse: the greatest number of violent storms for 154 years occurred during 2005, with 27 named storms and 16 hurricanes, of which three were Category 5 (the highest rating). The increase is put down to the side effects of a warming ocean.

Global warming is considered one of the most serious threats to the Caribbean. Climate modellers agree that the islands are likely to experience significant summer

drying by the middle of this century, creating wholesale drought. In addition, the land area is similar to that of the British Isles, but parcelled up into a multitude of islands, islets and cays – no more than 230,000 sq km (88,800 sq miles) in more than 4 million sq km (1½ million sq miles) of sea. It is not hard to imagine that a region with so much coastline and low-lying land, and which relies mainly on tourism for its economy, is highly vulnerable when sea levels are rising, as they are at the moment.

In the following chapters, we explore the principal islands and some of the surrounding mainland, especially the coastal strips and estuaries of Central and South America that border the Caribbean Sea. We also include the 'outer Caribbean' islands (with the exception of Bermuda, which is way out in the Atlantic Ocean) to the north of the Greater Antilles, and touch on the Florida Keys area that is very much influenced by the climate, ocean circulation and geology of the Caribbean and host to many of the same plants and animals. We travel clockwise around the Caribbean Sea, starting our journey on its largest island, Cuba, and in doing so we discover that this is a region with a chequered but intriguing natural and unnatural history.

ABOVE The swirl of clouds around the eye of a North Atlantic hurricane, as seen from space.

1

CUBA

ALTHOUGH CUBA forms part of the Greater Antilles, the island group that is the subject of chapter 2, its diversity of wildlife earns it a special mention of its own.

Limestone outcrops with clumps of patchy forest hanging precariously on vertical slopes, all bathed in an eerie morning mist – you would be forgiven for mistaking this scene for one in China or Japan. Columbus did. He landed on this, the largest island in the Caribbean, on Sunday 28 October 1492 and thought he was on the other side of the world. Those conspicuous outcrops or hummocks are known locally as *mogotes*, and they protrude from the floor of the Valle de Viñales, an area of fertile land, famed today for its tobacco. Ceibón or sacred silk-cotton trees *Bombacopsis cubensis* grow from the sheer cliff faces, along with piñón trees *Erythrina cubensis* and caiman oak *Tabebuia calcicola*. Huge pine forests, dominated by the Caribbean pine *Pinus caribaea,* clothe the surrounding mountains.

PREVIOUS PAGES Encircled by mountains, Viñales Valley has a microclimate that is perfect for growing tobacco.

ABOVE The limestone outcrops or mogotes of Viñales Valley are riddled with caves such as the Cueva del Panal.

The side of one of two named outcrops, the Dos Hermanas or Two Sisters, is decorated with Leovigildo González Morillo's giant 'Mural of Prehistory', 120 metres (400 feet) high and 180 metres (600 feet) wide, while deep inside the mogotes huge and complex cave systems contain the real thing. The largest is the Cueva del Santo Tomás, with a complex of caverns extending 45 km (28 miles) underground. In its depths are fossil skeletons of extinct ground sloths, including the black-bear-sized *Megalocnus*, evidence that at some point in the distant past the islands of the Greater Antilles were connected by a landbridge to the North American mainland.

In the cave's underground streams live three species of poorly sighted and almost colourless cavefish, including the 9-cm (3½ inch) long Cuban cusk-eel *Lucifuga subterraneus,* which feeds on aquatic cave isopods. The toothed cusk-eels *L. dentata* and *L. teresinarum* complete a trio unique to Cuba.

In fact, Cuba possesses the highest number of endemic species of any place in the Caribbean. Today, more than half its entire fauna and flora is found nowhere else in the world, and 40 per cent of all the animals encountered in recent biological surveys are completely new to science. The island is just 150 km (100 miles) to the south of Key West, Florida, but it might as well be a million miles away. Before Columbus arrived the entire island was forested, but despite deforestation and fragmentation during the past hundred years, Cuba under its current regime has been surprisingly successful at preserving its remaining forests and replanting anew. Since the revolution in 1959 that brought Fidel Castro to power, the island has been politically isolated; the lack of large-scale development coupled with the replanting of native forests has resulted in tracts of relatively healthy wilderness. Today, 15 per cent of the island is forested, and one-fifth of all the land area is protected as biosphere reserves or national parks, the largest proportion of any nation on Earth. As a consequence, 355 species of birds have survived here, 165 of them neotropical migrants that breed in North America and use Cuba as a stopover or overwintering area, mainly occupying the lowland, moist, semi-deciduous forests.

ABOVE The Caribbean reef shark was first described from a specimen that was caught off Cuba in 1876.

Cuba's first biosphere reserve was Sierra del Rosario, at the western end of the island. Created in 1985, its mountain slopes are cloaked in bastard mahogany *Batonia apetala*, with wood similar to genuine mahogany, and blue mahoe *Hibiscus elatus*, which has a durable wood used in cabinet making.

The island's 48 protected areas cover the greatest diversity of habitats in the West Indies, and some harbour record-breaking natural features. On the north coast, Buenavista Biosphere Reserve has spectacular karst scenery of limestone islands and cave systems. Within its boundaries, the island of Cayo Guillermo boasts the highest sand dune in the Caribbean – 15 metres (50 feet). In south-central Cuba, the Cueva San Martín Infierno contains the tallest stalagmite in the Americas – 67 metres (220 feet). In the southeast, the Sierra Maestra contains Turquino National Park, named after Pico Turquino, at 2005 metres (6578 feet) above sea level Cuba's highest peak and site of the first guerrilla stronghold during the revolution. At the western tip of the sierra, Desembarco del Granma National Park has some of the most impressive uplifted marine terraces and coastal cliffs in the western Atlantic. They stretch from 180 metres (600 feet) below sea level to 460 metres (1500 feet) above, and are home to the extremely rare blue-headed quail-dove *Starnoenas cyanocephala* and colourful painted snails, one species of which, *Polymita brocheri*, can be found in a small section of these terraces and nowhere else on Earth.

In the seas surrounding Cuba, marine life – sea turtles, sharks, dolphins and colourful reef fishes – is prolific, and there are 3735 km (2320 miles) of coastline in which to find it. Diving here is the sport of the élite, Castro having been a devotee, and one of the finest dive sites in Cuba is probably the remote Playa María la Gorda, a palm-lined pirate cove cluttered with ancient wrecks, cannonballs and huge shoals of snapper and parrotfish.

The island's 48 protected areas cover the greatest diversity of habitats in the West Indies and some harbour record-breaking natural features

The largest marine park in the Caribbean is the Jardines de la Reina or Queen's Gardens. Situated about 100 km (60 miles) to the south of Cuba's south-central coast, it has a chain of 250 coral cays and hundreds of islets that stretch for over 120 km (75 miles), claimed as possibly the third-largest barrier reef in the world. Commercial fishing is banned, so shoals of fish regularly attract eight species of

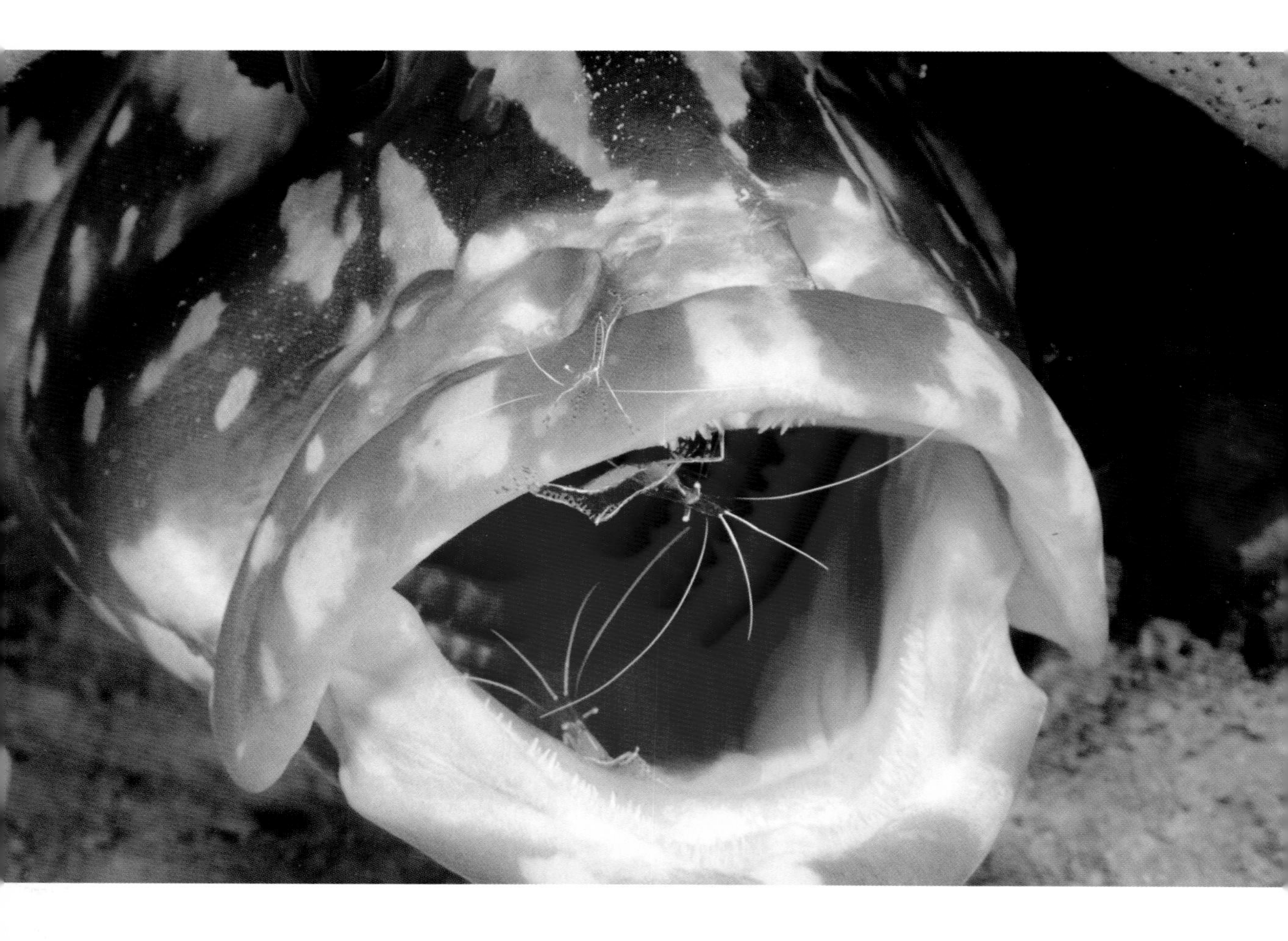

ABOVE The Nassau grouper *Epinephelus striatus*, here being cleaned of parasites by cleaner shrimps, is found throughout the Caribbean.

requiem sharks of the family Carcharhinidae, huge jewfish or goliath groupers *Epinephelus itajara*, tarpon *Megalops atlanticus* and red, black and yellow-mouth groupers *Epinephelus* spp.

The biggest protected area by far is the Ciénaga de Zapata Biosphere Reserve, 500,000 ha (1.25 million acres) of swamp, including salt- and freshwater marshes, mangroves and woodland, and covering a vast lowland area west of the Bay of Pigs. The Caribbean's equivalent of the Florida Everglades, it is recognized as a Wetland Area of International Importance.

Ciénaga de Zapata is set on a vast limestone plain and is a nursery for fish and a refuge for birds. Twenty-two of Cuba's 25 endemic birds live here, including the Zapata wren *Ferminia cerverai* and the Zapata rail *Cyanolimnas cerverai*. The Zapata sparrow *Torreornis inexpectata*, of which there are no more than a thousand left, and 90 per cent of the world population of Fernandina's flicker *Colaptes fernandinae* are also residents. Nature, though, has little respect for rarity. The flickers are regularly harassed by another local bird, the West Indian woodpecker *Melanerpes superciliaris*, which raids their nests and kills their chicks.

Zapata is also home to the world's smallest bird – indeed, the world's smallest warm-blooded vertebrate. Flitting among flowers and feeding on nectar at the forest edge, the bee hummingbird *Mellisuga helenae* is no bigger than 6 cm (2½ inches) long and weighs little more than 2 g ($\frac{1}{15}$ ounce). Its tiny wings have a span of 3.25 centimetres (⅓ inch), and they beat at over 80 times a second in flight and a staggering 200 times a second when hovering or mating. About 30 per cent of this tiny bird's body weight is wing muscles, and they propel it along at a top speed of 50 km/h (30 mph). As it zips about, it could easily be mistaken for a bee. Its heartbeat is the second fastest of any known animal and its body temperature, at 40 °C (104 °F), is the highest of all the birds. Its energy needs are such that it consumes half its total body weight in nectar and insects each day and drinks eight times its weight in water. In order to conserve energy at night, its body temperature drops to 30 °C (86 °F).

The bee hummingbird's nest is just 3 cm (1¼ inches) across, smaller than an eggcup; in it the female lays two pea-sized eggs. Local people consider this the bird of love and have christened it *zunzúncito*, meaning 'whispering' in Creole. It is found throughout the island, but has a patchy distribution due to the destruction or modification of much of its habitat.

Another treasure to be found here is the Cuban tody *Todus multicolor*, one of a genus of insect-eating birds that is unique to the Greater Antilles, with a separate species present on each island, except Hispaniola. When hunting, the tody perches

BELOW LEFT The noisy West Indian woodpecker is the scourge of flicker populations in the Zapata Swamp and quite common throughout Cuba.

BELOW The Cuban tody is thought to be the oldest surviving member of the tody family.

and scans for flying insects. If one comes by, it darts out in a flash of colour, grabs its prey and returns to its vantage point, where it beats the prey on the branch. It lives and hunts in wooded areas at forest margins.

Large tracts of the Zapata swamp are dominated by stands of sawgrass *Cladium jamaicense* and sugar-cane plume grass *Erianthus giganteus*, but the most distinctive plants are the palms. The royal palm *Roystonea regia* appears on Cuba's coat of arms, but it is just one of over 70 endemic species. The most important in Zapata is the sabal or cabbage palmetto *Sabal palmetto*, which harbours and feeds many birds and reptiles. The black-cowled oriole *Icterus prosthemelas* weaves its nest underneath the fronds, and the Cuban blackbird *Dives atroviolacea* and Greater Antillean grackle *Quiscalus niger* nest in the crown. Woodpeckers bore holes in the trunks, which in turn are the adopted homes of the Cuban pygmy owl *Glaucidium siju*, the extremely rare Cuban parakeet *Aratinga euops* and the Cuban trogon or tocorro *Priotelus temnurus*, a bird whose plumage is the same colour – red, blue and green – as the national flag and so has been adopted as the national bird.

OPPOSITE Christopher Columbus was the first European to see the West Indian manatee. His sailors mistook it for the mythical 'mermaid', a creature half fish, half woman.

The palm's small black fruits (drupes) are important for refuelling passing migrants such as Cape May warblers *Dendroica tigrina*, as well as being one of the favourite items on the dietary sheet of residents such as red-legged thrushes *Turdus plumbeus* and stripe-headed tanagers or western spindalis *Spindalis zena*.

The sabal palm can grow to 20 metres (65 feet) tall and live for a hundred years, thriving in relatively poor soils and able to bend in strong winds. Hurricanes, however, are not the major hazard. Local peasants who cut down palm groves to clear land for subsistence farming are the real threat to its survival. They use the wood to make charcoal for cooking and strip off the 2-metre (6½-foot) long leaves to thatch their huts. It is a tradition that Columbus observed and noted in his log in 1492. Today, as then, it takes at least a thousand leaves to thatch a single hut, or the leaves from 30 trees.

In the flooded areas, water hyacinths *Eichhornia crassipes*, fragrant water lilies *Nymphaea odorata* and water lettuce *Pistia stratiotes* are common, and in deeper waters, beds of aquatic vegetation, such as sea grass *Thalassia* spp., are food for West Indian manatees *Trichechus manatus*. These 3–4-metre (10–12-foot) long, slow-swimming, aquatic mammals were once common in the estuaries of Cuba's rivers, but contamination, damming and deforestation have seen them move to more protected coastal sites, like the relatively undisturbed seaward edge of Zapata.

The first written record of the manatee was probably the three 'mermaids' observed by Columbus and his crew in 1493, three months after his arrival in the West Indies. The ship's log for 9 January states that 'they were not as beautiful as they are painted, though they have something of a human face.' The crews of later ships, however, related

'They were not as beautiful as they are painted, though they have something of a human face'

to the manatee in quite a different way – they ate them. The Maya had a way of drying manatee meat that was called *buccan*, and the pirates who raided Spanish treasure ships in the Caribbean relied so much on it in their sparse diet that they became known as 'buccaneers'.

The manatees share their watery home with the manjuarí or Cuban gar *Atractosteus tristoechus*, a 'primitive' bony fish, and the only garfish to have adopted a brackish-water existence that can take it into the fully marine environment. Exceptional individuals can grow up to 3 metres (12 feet) long, although 2 metres (6 ½ feet) is the average nowadays. In shape, the manjuarí resembles a long, thin crocodile with a long snout and needle-like teeth, and its languid lifestyle belies an aggressive beast. It lurks among underwater plant life, waiting for prey to drift by, and has even been known to tackle a passing crocodile. While its flesh is edible, its eggs are poisonous, a fact that might have helped it to survive for over 100 million years, an achievement validating its label of 'living fossil'.

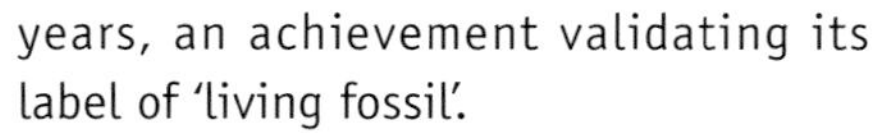

In the shallower waters, one small part of the Zapata Reserve is home to another ancient predator – the Cuban crocodile *Crocodylus rhombifer*. More than half of the world's entire population survives here. The crocs average about 2 metres (6–7 feet) long, so are relatively small in crocodile terms, but what they lack in size they make up for in attitude. These feisty beasts are considered the most aggressive of all the crocodilians. They have been known to propel themselves right out of the water in order to grab animals from overhanging branches, such as the hutia – a large and endangered tree-climbing rodent.

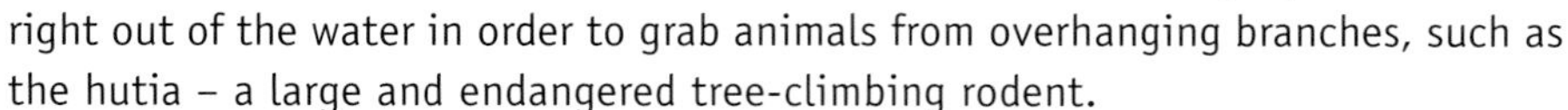

Hutias or *jutía* are confined to the islands of the Greater Antilles, and Cuba has ten species. Some are rare, but others are abundant and hunted for the pot using dogs. They are flushed out from dense forests, where they not only forage on the ground, but also clamber about in the trees and in caves, where they grip the slippery rocks with their claws. They have a varied diet, from leaves and bark to insects and lizards, their three-compartment stomach able to digest just about anything. They often hide in the hollow trunks of dead sabal palms, a shelter used also by nesting parrots.

The species of hutia surviving in the Zapata swamps include the Cuban hutia or *jutía conga, Capromys pilorides,* and the smaller dwarf hutia or *jutía enana, Mesocapromys nanus*. The latter was once common throughout Cuba, but today occurs only on dry islets within the swamps. Its ability to increase its population

LEFT A hutia in caves on the Peninsula de Guanahacabibes at the western tip of Cuba.

significantly is hampered not only by introduced mongooses, but also by the fact that females, in a very un-rodent-like manner, have just one youngster at a time.

Other hutias are found on small, remote islands offshore. Cabrera's hutia *Mesocapromys angelcabrerai*, for instance, lives in the mangrove forests of Cayos de Ana Maria, off the coast of south-central Cuba, where it builds circular, 1-metre (3-foot) diameter, communal nests of branches and leaves up in the trees. All hutias have been given 'critically endangered' status by the International Union for the Conservation of Nature and Natural Resources (IUCN), but in some parts of Cuba and its islands they are still hanging on.

BELOW The Cuban crocodile is adept at walking and leaping, as well as swimming.

The hutia's natural enemy is also considered to be the world's most threatened crocodile species. No more than 6000 remain, most living in the Zapata swamp or in the Lanier swampland on the Isla de la Juventud (Youth), also known as the Isla de los Pinos (Pines). The crocodiles were once found on several other Caribbean islands, but they were hunted for their meat and skins, and their numbers declined rapidly. Today, captive-breeding facilities – one on the main island and another on Juventud – are successfully rearing the next generations, but there is a danger in the wild that they cross-breed with American crocodiles *Crocodylus acutus*, another rare species. A discrete population of this larger and more docile species is bred and reared under guard at the Monte Cabaniguán Wildlife Refuge, in the southeast of Cuba. On occasions, there are so many females nesting at any one time, in a small and restricted area, that the late shift digs up the eggs deposited by earlier arrivals.

Small is certainly beautiful in Cuba. Joining the smallest crocodile and smallest bird is a miniature amphibian – reportedly the northern hemisphere's smallest species of frog, *Eleutherodactylus iberia*. At 1 cm (less than ½ inch) long, this tiny black frog with longitudinal orange stripes can sit comfortably on a small coin, and ties with Brazil's gold frog *Psyllophryne didactyla* in being not only the world's smallest frog but also its smallest living tetrapod (four-limbed vertebrate). Cuban scientist Albert Estrada and S. Blair Hedges of Pennsylvania State University discovered it in 1996; they found it living among the leaf litter and under ferns in humid rainforest at an altitude of 600 metres (2000 feet) on the western slopes of the Monte Iberia tableland in eastern Cuba. There are only two known populations, the other being at sea level near Nibujón.

It is likely that many more new species of frogs, lizards and a host of other animal groups are still to be found, but no sooner are they discovered than they are placed on the critical list. There are probably no more than 250 individuals of one tiny frog, *Eleutherodactylus symingtoni*, living in caves and rock shelters in just six locations in the montane forests of western Cuba, for example. The population is thought to have dropped by 80 per cent in the last ten years, possibly due to disease.

However, conservation in Cuba is not an entirely depressing subject. On the banks of the Río Máximo in the north of Camagüey province is the largest-known congregation of Caribbean flamingos *Phoenicopterus ruber ruber* in the world. The birds fly in to this remote and inhospitable region to nest and, thanks to successful conservation measures, about 40,000 nests are occupied each year. The Caribbean species is the second-largest flamingo (after the greater, *P. roseus*, to which is closely related), growing to about 1.5 metres (5 feet) tall. It is also the most brightly coloured of all the flamingos, with vivid coral-red plumage. The pigment is obtained in their diet of algae, fly larvae and brine shrimps that they sift from the water and mud, using a filtering method much like baleen whales.

The Isla de la Juventud is the site of another success story – the resurrection of the Cuban parrot or *cotorra cubana, Amazona leucocephala*. Here, in Los Indios Ecological Reserve, an army of volunteers, including schoolchildren, keeps an eye on the nesting birds; at the turn of the millennium 1700 breeding parrots were counted, which means that the population has tripled in just over a decade.

Cuba's plant life is just as special as its animals. Of the 7000 known plant species, nearly half are endemic to the island and, as with the animals, many are extremely rare. Notable among these are *Pinguicola lignicola*, the world's only carnivorous epiphyte, and *Solandra grandiflora*, which has a flower 10 cm (4 inches) across at the calyx and 30 cm (12 inches) at the corolla, making it one of Cuba's largest blooms. Hundreds of species of orchid flourish in the wilderness areas, including the diminutive *Pleurothallis shaferi* – another of Cuba's dwarfs at just 1 cm (less than ½ inch) across.

Curiously, however, Cuba's national flower is not a native. The white butterfly *Hedychium coronarium* is not a flying insect but a delicate white-flowered jasmine from Vietnam. Its elevated status over local flowers came after it was adopted by the women who took part in the wars for independence in the nineteenth century. They wore it in their hair or pinned to veils and shawls as a secret sign of their allegiance to the liberation army, carrying messages hidden in the flowers between isolated guerrilla groups. Today, the white butterfly is often included in a bride's bouquet.

Rainforests occur at the eastern end of the island. The Cuchillas del Toa Biosphere Reserve is here and covers the mountainous region of Sagua-Baracoa in the Alejandro de Humboldt National Park. It contains habitats that range from cloud forest to coral reef, but more importantly it has the largest tropical, moist forest remaining in the Caribbean. Here, 928 endemic species have been recorded, including such botanical jewels as the very rare and primitive *Podocarpus angustifolius*, the small, dome-shaped melon-cactus or dwarf Turk's cap *Melocactus matanzanus;* and Moa's dracaena *Dracaena cubensis*, which is adapted to living in the poor soils associated with magnesium-rich serpentine rock formations and grows nowhere else.

Among the birds to be found here is the royal carpenter or ivory-billed woodpecker *Campephilus principalis*, which was thought to be extinct until a confirmed sighting in the Sierra de Moa mountains of eastern Cuba; it has also turned up in the USA, so it seems that it is still hanging on. Other residents are the caguarero sparrow hawk or Cuban kite *Chondrohierax wilsoni,* which feeds on snails, including the striped, multicoloured land snails *Polymita* spp., and the almiqui, known otherwise as the Cuban solenodon *Solenodon cubanus*, one of two species that live only on Cuba and Hispaniola.

The nocturnal solenodon is a giant shrew-like creature weighing about 1 kg (2¼ lb). It inhabits dense, humid forests, brush and plantations, where it snuffles about, rooting out insects and spiders hidden amongst the leaf litter. It will also tear at rotting logs with its long and stout front claws, probing with its elongated snout for the grubs living inside. Glands above its teeth secrete a poison that subdues its prey.

The solenodon is also unique among insectivores in adopting 'teat transport' as a means of carrying babies around while mother is foraging. Initially the offspring are dragged along, but eventually they run alongside her, still hanging on to her teats until they are about two months old.

During the day, the solenodon hides in rocky clefts, in hollow trees and in burrows that it can dig for itself, but it is far from safe. The introduction of the Burmese mongoose *Herpestes javanicus* in the 1870s to control rats in sugar-cane plantations is thought to have been responsible for the first phase of its drastic decline, although in recent years it has been hunted to near extinction by feral cats and dogs. In fact, in the early 1970s it was thought to be extinct, but it has since been rediscovered in several places, especially in the

The white butterfly *Hedychium coronarium* is not a flying insect but a delicate white-flowered jasmine

LEFT A flock of Caribbean flamingos on the banks of Cuba's Rio Maximo.

ABOVE One of the few places in the world to see the exceedingly rare ivory-billed woodpecker is the Cuchillas del Toa Biosphere Reserve in the northeast of Cuba.

The pigment is obtained in their diet of algae, fly larvae and brine shrimps that they sift from the water and mud, using a filtering method much like baleen whales

Holquin Province at the eastern end of Cuba. Even here, though, it is rarely seen, only one having been spotted in the past three years.

Some of the caves in mountains at this end of the island are approached over rocks so sharp they can shred hiking boots, but their principal inhabitants move in and out with considerably more ease: they fly, for they are bats – 12 species, including one of the world's smallest, the butterfly bat *Natalus lepidus*, a miniature insect-eater with a wingspan of just 13 cm (5 inches). These bats roost in caves both here and in the middle of the country, where they sometimes gather in such large numbers, they raise the temperature of their 'hot caves' to 40 °C (104 °F) or more. Wherever they are found, they follow the same daily ritual.

At half-past seven each evening, all the bats leave their caves to go hunting or foraging in the surrounding forests. They set out in an orderly fashion, the largest first. Some fail to fly far, though, for waiting in the trees near the cave entrances are Cuban boas or *majá de Santa María, Epicrates angulifer*. They anchor their bodies on branches, their mouths ready to strike at the aerial procession. Should a snake detect a passing bat, it grabs it in its jaws and then, in true boa fashion, squeezes out all signs of life. It then swallows its prey whole.

Cuba's most noticeable wildlife spectacle, however, is not hidden in a remote forest but out in the open for all to see. All along the southern coast during March and April, the coast roads are jammed with millions of red, yellow and orange land crabs *Gecarcinus* spp. For most of the year they live in the depths of the moist forest to avoid desiccation, but during the rainy season, in spring, they migrate inexorably to the coast, the journey taking them through the forest, across farms and even over highways to spawn at the edge of the sea. Breeding complete, they head back *en masse* to the forest. It is an annual hazard for crabs and motorists alike. The crabs are squashed as they cross the road, but they get their own back: their hard, outer skeleton punctures tyres – hundreds of tyres – every season. On 17 April 2001, their annual jaunt had unexpected consequences when the press entourage accompanying US dignitaries attending the fortieth anniversary of the unsuccessful invasion of the Bay of Pigs was left on the roadside. Crab debris had punctured the tyres of all their cars.

For US citizens to be here at all is a phenomenon, for Cuba was a closed shop for almost 40 years. Today, however, the country is open to visitors, with the inevitable consequence of escalating development. Nevertheless, there is a will to protect the wilderness, especially the less populated mountain areas and coastline that have yet to be denuded. Groups such as WWF Canada, Bird Studies Canada and the Canadian Wildlife Service work with the Cuban Institute of Ecology and Systematics (direct co-operation between Cuban and US organizations is difficult because of a trade embargo); and there are sufficient local people proud of their wilderness areas and their wildlife to press continually for a wild future.

BELOW The Bay of Pigs is one location where millions of Cuban land crabs leave the forest and move to the seashore to spawn each spring.

And, ironically, the US military base at Guantánamo Bay (known as 'Gitmo'), which has had so much negative press in recent years, is a hotbed of conservation. Naval security forces on the base take pride in their 'honorary' wildlife protection duties, and of the 91 sq km (35 sq miles) of land in this unusual US territory – the rest of its 184 sq km (71 sq miles) is sea and coral reefs – much remains as virgin forest, the richest vegetation of its kind in the entire West Indies; and that includes the minefield that separates it from the rest of Cuba. Only wildlife is allowed to enter the uninhabited areas. The result is that some of the animals living here grow larger than their counterparts on the rest of the island. Rock or ground iguanas *Cyclura nubila*, for example, grow to a length of 1.5 metres (5 feet), much larger than those elsewhere. Apart from the four new wind generators that provide about 30 per cent of the energy needs of the base, the unoccupied area is almost as it was when Columbus landed over 500 years ago.

2

GREATER ANTILLES

AS WE SAW in the introduction, the Greater Antilles comprise the four largest islands in the Caribbean – Cuba, Jamaica, Hispaniola (which is divided politically into Haiti and the Dominican Republic) and Puerto Rico – together with smaller island groups, such as the Cayman Islands.

The Caymans are an overseas territory of the United Kingdom, with three main islands – the largest, Grand Cayman, separated by 145 km (90 miles) of sea from Cayman Brac and Little Cayman. They lie 240 km (150 miles) south of Cuba, just a one-hour flight from Miami.

The Caymans were discovered quite by accident by Columbus on his fourth and last voyage to the West Indies in 1503. He was on his way from Panama to Hispaniola when strong winds blew his ship off course, and he landed here on 10 May. He was followed 83 years later by Sir Francis Drake, the first Englishman to set foot on these islands. It was Drake who gave them their current name, which is derived from the Taíno word *caiman*, meaning 'crocodile'. Today, the Caymans are noted for their offshore banking facilities and beach tourism, especially at the famous Seven Mile Beach, but there is an incredible amount of wildlife packed into this small area too.

Geologically speaking the three islands are large blocks of coral that grew on the peaks of a western extension of Cuba's Sierra Maestra, which were submerged after the last Ice Age. They are remarkably flat, the only visible highland being the Bluff, a limestone outcrop on Cayman Brac, which rises to a height of 43 metres (140 feet) above sea level. It is the feature that gave the island its name, for *brac* is the Gaelic name for bluff.

Grand Cayman's highest point is just 18 metres (60 feet) above the sea, but its rocky terrain covered in dense undergrowth makes much of the island inaccessible, especially in the east. There is, however, one area in which visitors can get to the heart of this small wilderness. Mastic Reserve is the largest area of old growth, subtropical, semi-deciduous dry forest left on the island, and it contains plants and animals unique to the Caymans, including the yellow mastic tree *Mastichodendron (Sideroxylon) foetidissimum*. A 3-km (2-mile) long trail, restored from one constructed over a hundred years ago, was opened in 1995. It includes the 170-metre (550-foot) long 'Mastic Bridge' causeway of mahogany logs and beach rocks that has been built through black mangrove wetland.

Stands of palms and rare trees such as cedar and mahogany are scattered throughout the reserve, hiding native parrots *Amazona leucocephala caymanensis* (Cayman Brac has its own subspecies – *A. leucocephala hesterna*), West Indian woodpeckers *Melanerpes superciliaris* and Caribbean doves *Leptotila jamaicensis*. On the ground are ranks of tiny soldier crabs *Coenobita clypeatus*, a form of land-based hermit crab that lives among the leaf litter below sea grape trees *Coccoloba uvifera*. Staying on the trail here is recommended, as poisonous plants such as maiden plum *Comocladia dentata*, which causes serious skin reactions, and the lady hair

PREVIOUS PAGES The loggerhead turtle is a frequent visitor to Caribbean beaches, where females deposit their eggs in the sand.

OPPOSITE Tree ferns in the sub-tropical forest of the Caribbean National Park in the rugged Sierra de Luquillo, Puerto Rico.

Malpighia cubensis, whose fine hairs sting and irritate the skin like fibreglass, are lurking in the undergrowth for the unwary.

Each June the epiphytic banana orchid *Schomburgkia thomsoniana* – the national flower of the Cayman Islands – blooms at the trailside. In South America, this plant is 'guarded' by stinging ants, which make their home in its pseudobulb and feed on sweet nectar from the flower spike. The Cayman orchids possess the pseudobulb and spike but no ants; instead, anole lizards climb up to the flowers and lick the nectar.

OPPOSITE The Cayman blue iguana is found in the Cayman Islands and nowhere else in the world.

The largest protected area on Grand Cayman is the Salina Reserve, located in the northeast of the island. Although a wetland area of sedge and buttonwood, it dries out in the summer, leaving a covering of desiccated algae on the surface, much like the salt crust of saltpans elsewhere in the West Indies – hence the name. There are no trails here, only dense vegetation, so the area is relatively undisturbed. Brazilian free-tailed bats *Tadarida brasiliensis*, Jamaican fruit-eating bats *Artibeus jamaicensis* and big-eared bats *Macrotus waterhousii* occupy a wooded ridge, and bald pates or white-crowned pigeons *Patagioenas leucocephala* nest in the dry forests. There is even a small pink-flowering herb *Agalinis kingsii*, found nowhere else on Earth.

One of the crowning glories of the reserve is the controlled-release programme of captive Grand Cayman blue iguanas *Cyclura nubila lewisi*, a distinct subspecies from the rock iguanas on Little Cayman and Cayman Brac. The blue coloration is more evident in the males and appears during the heat of the day. In the early morning the lizards are dark grey overall, a good colour for absorbing the heat from the sun. To prevent them overheating, their undersides turn a distinct powder blue that absorbs less heat. They are vegetarians with a strong bite and can grow to 1.5 metres (5 feet) long; in fact, they do not stop growing throughout their life. At one time, they deposited their eggs in the sand on the island's many beaches, but the arrival of people and domestic animals forced them to move to less accessible places inland. Only on Little Cayman do the rock iguanas still deposit their eggs at the edge of the beach. The National Trust for the Cayman Islands has a programme to save the remaining lizards and breed new ones for release. The only problem is fire, caused mainly by lightning strikes. In some years, all the sedge wetland can burn, but when the rains come the vegetation miraculously sprouts once again.

To prevent overheating, the iguanas' undersides turn a distinct powder blue that absorbs less heat

Other notable wildlife landmarks on Grand Cayman include the Governor Michael Gore Bird Sanctuary, whose freshwater pond attracts resident and migrant birds even when the rest of the island is dry, and Meagre Bay Pond, which is a magnet for waterfowl. Willie Ebanks' Farm, on the north side of Malportas Pond, is a haven for the endangered West Indian whistling duck *Dendrocygna arborea*. As its scientific name indicates, it is more closely related to swans and geese, and spends a lot of time in trees. Its common name refers to its haunting four-to-five-syllable whistle.

An important tree species is the endemic silver thatch palm *Coccothrinax proctorii,* the national tree of the Cayman Islands. Its slender trunk grows to 9 metres (30 feet) tall, and it produces a profusion of small white flowers, which turn into berries that change colour from green to red to black. The leaves are light green on the topside and silver underneath, and they seem to shine in moonlight. They have been used to thatch houses, to weave hats, baskets and fans, and to provide thongs for 'wompers', a form of sandal. More importantly, the palm's dried leaf is not damaged by sea water, so it was used to make rope.

West Bay, in the northwest Grand Cayman, has the distinction of being not only the location of a large sea-turtle farm, but also the site of jagged, black rocks, the result of bacteria having attacked the white limestone. It has become known affectionately as 'Hell', after a local official declared that this must be what hell looked like; tourists can now send home a postcard from Hell.

The other Cayman islands may be smaller, but they are no less prolific. Cayman Brac has the Brac Parrot Reserve, centred on the Bluff and home to 400 Cayman Brac parrots (also known as 'stealth parrots' on account of their near perfect camouflage), a species that has the smallest range of any known Amazon parrot. The reserve is a curious place where cactuses, *Pilosocereus* spp., play host to epiphytes, such as banana orchids and bromeliads *Tillandsia* spp., and share living space with hardwoods and century plants *Agave sobolifera*.

ABOVE LEFT About 20,000 red-footed boobies nest on Little Cayman, the largest breeding colony in the Western Hemisphere.

ABOVE The male magnificent frigate bird inflates his bright red pouch to impress the females.

Little Cayman's Booby Pond Nature Reserve claims the largest colony of red-footed boobies *Sula sula* in the western hemisphere. The birds nest in the mangroves and feed at sea, sometimes reaching the coast of Cuba or Jamaica. Each evening, however, when they return to their nest sites, they are confronted by pirates. Magnificent frigatebirds *Fregata magnificens* engage them in aerial dogfights and force them to disgorge their food. The boobies do not take this lightly, though. Returning birds join with others of their kind and form a huge spiralling column high into the sky. Gradually, they peel off and in a controlled freefall 10,000 birds dive down to their nest in the mangroves, leaving stragglers to contend with the raiders offshore.

Surrounded by shallow, warm, crystal-clear waters, all the Cayman islands are fringed by mangroves, including Grand Cayman's Central Mangrove Wetland or Central Swamp, considered by some to be the Cayman's ecological heart. As with small islands everywhere, the focus for most visitors is the sea itself. All three islands have élite diving sites, but Grand Cayman has the most famous marine residents – the southern stingrays *Dasyatus americana*. A series of shallow sand bars a short boat ride from the northern end of the island is known as 'Stingray City', and here visitors equipped with nothing more than a mask and snorkel can feed and pet squadrons of seemingly 'friendly' rays.

There are also magnificent coral reefs, with barrel sponges, reef fishes, barracudas, sharks and sea turtles in exceptionally clear, warm water without strong currents, but experienced divers head for the edge of the reef where the water suddenly becomes deep, the so-called Cayman Wall. The underwater northeast wall of Grand Cayman has the Valley of Dolls, probably the most picturesque site with a 2.5-metre (8-foot) tall feather plume *Pseudopterogorgia* spp., black corals and 17 massive barrel sponges *Xestospongia muta*. These bulky sponges pump 10,000 times their own volume of water each day, yet for every tonne of water strained they recover no more than 30 grams (an ounce) of food.

The southeast wall has the Three Sisters with huge pillars of coral 18 metres (60 feet) in diameter, Scuba Bowl with a massive mushroom-shaped coral formation resembling a church and Jack McKenney's Canyon (named after the late underwater film-maker), the deepest drop-off in the northern hemisphere – 7500 metres (25,000 feet) straight down.

The *Kirk Pride*, a cargo ship that sank in 1976, rests on a ledge about 270 metres (880 feet) down the Cayman Wall, providing anchorage for fine, spindly corals and pale, flower-like stalked crinoids or sea lilies up to 1.2 metres (4 feet) long. Curious dull-red stalactites, known as rusticles, hang from the steel plates.

BELOW A southern stingray lifts off the sea bed where it was buried in the sand with just its barbed tail exposed.

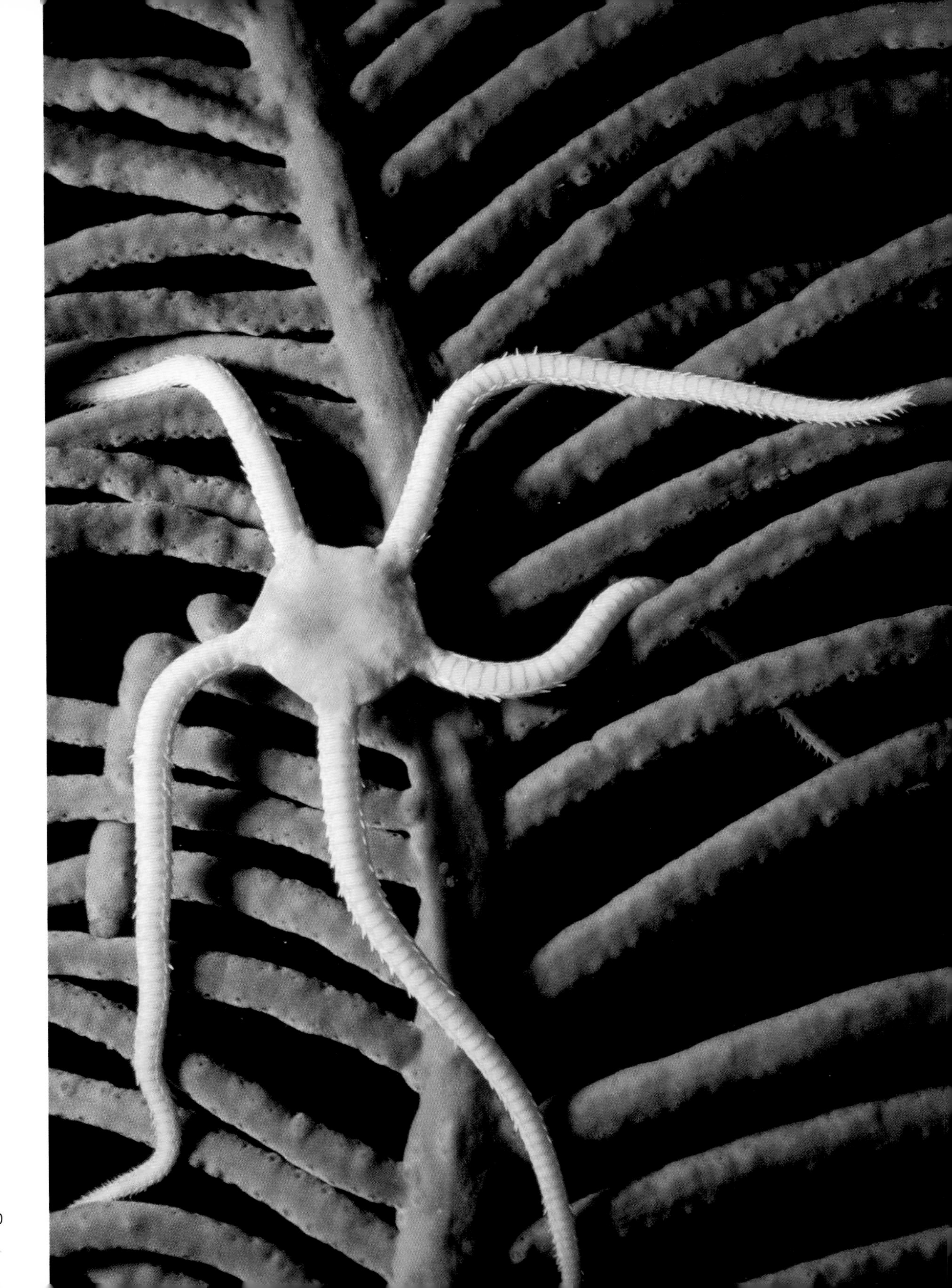

They are formed by the bacteria that are eating away the ship's hull and have an internal structure similar to tree rings. When touched they just fall away as dust, the fate eventually of the ship itself.

The proximity to deep water is due to a relatively narrow trough in the sea floor. It has a southwest-northeast axis and separates the Cayman Islands from Jamaica. This is the 1700-km (1050-mile) long Cayman Trench, where the maximum depth is 7686 metres (25,216 feet), the deepest point in the Caribbean Sea. Scientists have yet to unlock the trench's secrets, the exception being a spell of shark research that has revealed large schools of Cuban dogfish *Squalus cubensis* patrolling the depths, along with gulper sharks *Centrophorus granulosus*, the rare blotched catshark *Scyliorhinus meadi*, sharp-nosed seven-gilled sharks *Heptranchias perlo* and big-eyed six-gilled sharks *Hexacanthus nakamurai*.

On the other side of the 100-km (60-mile) wide trench is the third largest island in the Caribbean – Jamaica – visited by Columbus in 1494. It is ringed by a narrow coastal plain, indented with bays and sand beaches to form a coastline over 1000 km (600 miles) long. One of the most dramatic beaches is the crescent-shaped Long Bay, in the northeast of the island, with rose-coloured sands lapped by a deep turquoise sea.

The interior is of limestone, with caves containing the petroglyphs (stone carvings) made by the indigenous Arawak people. One area important for wildlife here is known locally as Cockpit Country, a limestone plateau about 600 metres (2000 feet) above sea level, in the northwest of the island. Its surface of limestone fissures and conical hummocks, divided by ravines and pockmarked with sinkholes and caves, earned it its name in the seventeenth century because it reminded observers of cock-fighting arenas.

The caves have significant palaeontological interest for it was here, in Long Mile Cave, near Windsor, in 1920, that the fossils of an ancient Antillean monkey *Xenothrix mcgregori* were discovered – evidence that Jamaica once supported a native population of primates related to those in South America. The fossils were found alongside those of the black rat, that had been brought to the islands by the Europeans, indicating that they were relatively modern. The European invaders must have hunted the monkey for food, and it is thought to have disappeared in the 1750s. This means that Jamaica is the only Caribbean island known to have had its own primate species, and that the Jamaican monkey is the only monkey in the world known to have gone extinct during the past 500 years.

Like many places in the Caribbean, today the area is home to other seriously endangered animals. It is the last stronghold of the black-billed parrot *Amazona agilis*, the smallest of the Amazon parrots. It arrived most probably by island-hopping from Honduras and today can be seen flying in small family groups in Cockpit Country. It differs from Jamaica's other endemic Amazon, the yellow-billed parrot *A. collaria*, in having a faster wing beat and a higher-pitched call. The yellow-billed is larger and more closely related to the Cuban parrot *A. leucocephala*, and is

OPPOSITE The gaudy brittlestar *Ophioderma ensiferum*, a recently discovered species, emerges from its daytime hiding place and crawls over a sea plume *Pseudopterogorgia*.

found alongside its more distant relative only in this Cockpit region of Jamaica.

The parrots nest in the hollows in trees, and they have over 46 species from which to choose, including the breadnut *Brosimum alicastrum*, black lancewood *Oxandra lanceolata*, timber sweetwood *Nectandra sanguinea* and silk cotton *Ceiba pentandra*. These trees, however, are not endemic. The flora unique to Cockpit Country is confined to herbs, shrubs, bromeliads, wild yams *Dioscorea villosa* (once the only source of the main ingredient in contraceptive pills) and six species of *Lepanthes* orchids.

Endemic invertebrates are represented by a multitude of land crabs that inhabit streams, caves and sinkholes. One species, *Metopaulias depressus,* has taken to depositing its young in the small pool of water held by bromeliads. The females of this species show extraordinary maternal behaviour, defending their offspring against an endemic predatory damselfly, *Diceratobasis macrogaster,* and even bringing pieces of snail shell to buffer the water made acidic from decomposing vegetation. The increased calcium content from the shells helps during the moulting and enlargement of the exoskeleton that occurs when the little crabs are growing. Frogs also use bromeliad tanks. Several female hylid frogs deposit their eggs here and, due to the lack of food, they provision their tadpoles with their own unfertilized eggs.

At night, during spring and summer, the hillsides glow with the lights of fireflies, including the endemic *Photinus synchronans* whose individuals flash synchronously. By day, higher elevations are the domain of the giant swallowtail *Pterourus (Papilio) homerus*. With a wingspan of 15 cm (6 inches), this is the second largest butterfly in the world. It is also found in the Blue Mountains and John Crow Mountains at the eastern end of the island, the former being the site of Jamaica's highest mountain, Blue Mountain Peak, 2256 metres (7402 feet) above sea level.

Another key conservation area is Black River's Great Morass, 325 sq km (125 sq miles) of swampland at the convergence of five rivers, through which Jamaica's longest river passes. The banks of the Black River, named for its mahogany-coloured water, are lined either by red mangroves or by salt grass, and hidden among the vegetation are myriad water and marsh birds. Fat snook *Centropomus parallelus* and tarpon snook *C. pectinatus* lurk below the surface, but the undisputed top predator is the American crocodile *Crocodylus acutus*. Since hunting was banned in the 1970s, about 300 survive here, a respectable population but still vulnerable.

The undisputed top predator is the American crocodile *Crocodylus acutus*

Jamaica's largest remaining mangrove system is included in the Portland Bight Protected Area. When Columbus came here on 18 August 1494, he named it Bahía de las Vacas, meaning 'Bay of Cows', on account of its huge population of West Indian manatees. Today, numbers have dwindled to such an extent that only one mother and her calf survive.

The next major landmass to the east is Hispaniola, divided into Haiti, which occupies the western third of the island, and the Dominican Republic. Here the geography is probably the most diverse in the entire Caribbean. Five mountain ranges cross the island, creating some sites where there is no rain and others where it never stops.

Haiti is a bit of an ecological disaster, with most of its native forest removed and many of its coral reefs silting up from debris washed down from the denuded slopes. Wild places are found only where the land is least accessible, but there are a few. High-altitude pines, some 45 metres (148 feet) high and 2 metres (6½ feet) in diameter, occur in cloud forests where 4000 mm (150 inches) of rain falls each year. There are also undisturbed mangroves in the more remote coastal locations.

One of Haiti's last wilderness areas is the Parc National Pic Macaya, about 180 km (110 miles) west of Port-au-Prince. Here, black-capped petrels *Pterodroma hasitata* nest, and 40 per cent of Hispaniola's orchid species found in a 20-sq-km (7-sq-mile) patch of surviving broadleaf forest.

BELOW American crocodile is distinguished from the alligator by the fourth tooth on either side of the lower jaw being visible when the mouth is closed.

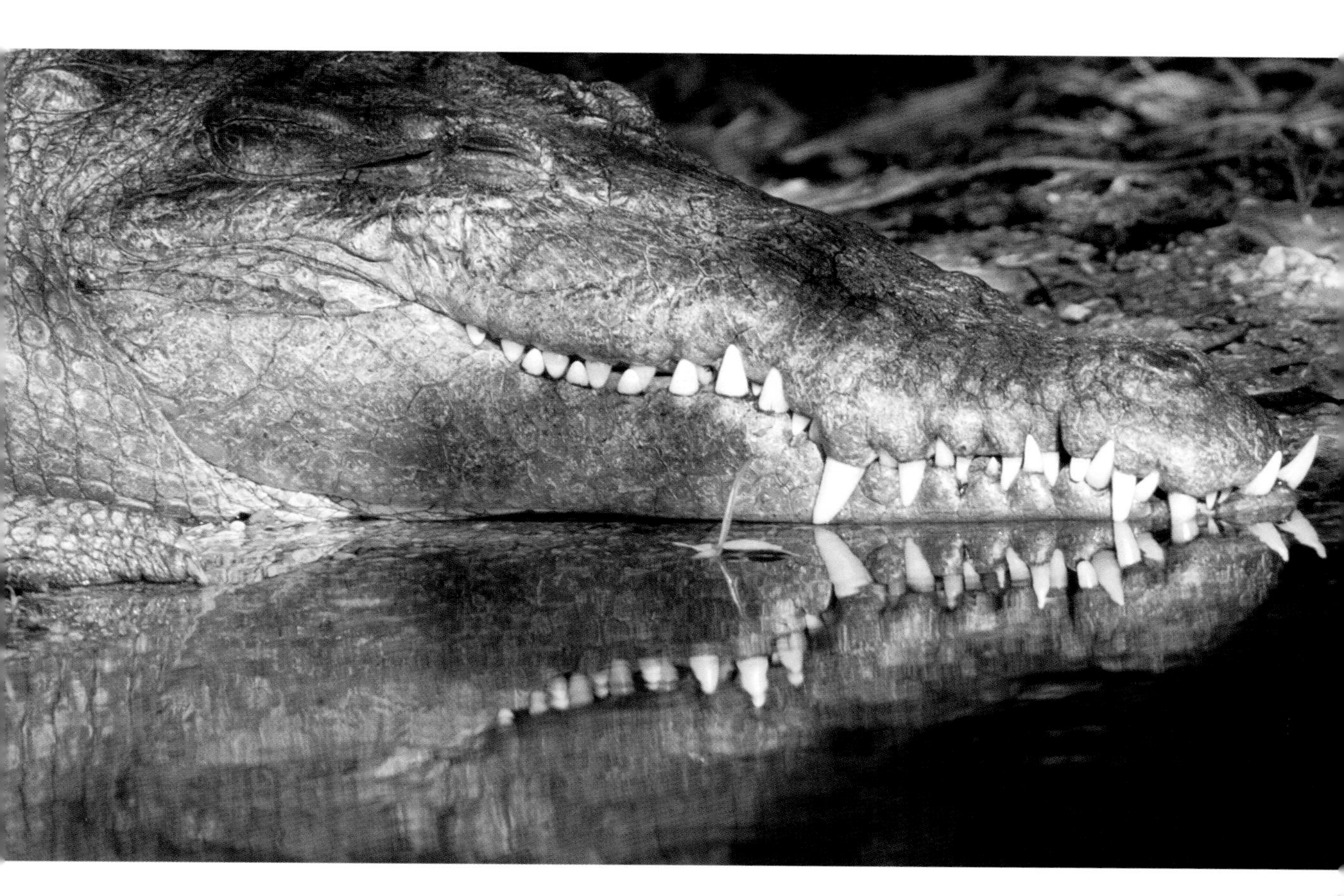

ABOVE The powdery white sands and swaying palms of Bacardi Beach at Cayo Levantado, a small island in the Dominican Republic.

One other place that has remained in pristine condition is the tiny, uninhabited, oyster-shaped island of Navassa, about 55 km (35 miles) west of Haiti's Tiburón Peninsula. Politically it is not part of Haiti, but a territory currently governed by the USA. Consisting of a raised coral and limestone platform ringed by vertical white cliffs up to 14 metres (45 feet) high, Navassa is covered by scrub and was once a guano-mining centre, the product of large seabird colonies, including a 5000-strong population of red-footed boobies.

Navassa has several endemic reptile species, the most visible being a green anole *Anolis longiceps* and a dwarf gecko *Sphaerodactylus becki*. Many more of its treasures are in the sea. US scientists have discovered that Navassa's reefs and underwater life have remained untouched and undisturbed. Already several new species of coral reef fish have been discovered (as well as new species of spiders on land). A US exclusion order, declaring the area a National Ocean Wilderness, will ensure that it remains undamaged.

On the main island of Hispaniola, the Dominican Republic has the highest and lowest locations in the region. In the Cordillera Central, the greatest of Hispaniola's mountain ranges, Pico Duarte – at 3098 metres (10,164 feet) – is 2 metres (6½ feet) higher than La Pelona, its twin peak, and the highest point in the entire Caribbean. Straddling the border between Haiti and the Dominican Republic is a rift valley that stretches for 127 km (79 miles). Like its African counterpart, the rift is partly below sea level and contains a string of salt lakes, including Lake Azuéi in Haiti and Lake Enriquillo in the Dominican Republic, the latter, at 44 metres (144 feet) below sea level, being the lowest point on land in the Caribbean.

Enriquillo is the only salt lake in the world to have crocodiles; and it boasts the world's largest single population of American crocodiles. The lake contains three islands, the largest being Isla Cabritos, which has a national park famous not only for the crocs but also for its flamingos *Phoenicopterus ruber,* which arrive here in huge numbers to feed on brine shrimps.

The land surrounding the lake is also significant. The cactus and thornbush scrub of the Enriquillo Basin is home to the endangered rhinoceros iguana *Cyclura cornuta*. The males of this species are especially ferocious-looking, with three small horns on the snout and a pad-like helmet on the top of the head. In reality, though, they are timid vegetarians, feeding on leaves and berries.

Palm trees, especially royal palms *Roystonea* spp., are home to the communally nesting palm chats *Dulus dominicus*, which pack 30 breeding pairs into a nest 1 metre (3¼ feet) wide and up to 3 metres (10 feet) high. Hispaniola's only endemic raptor, Ridgway's hawk *Buteo ridgwayi* (actually a buzzard, not a hawk), is an occasional visitor, although most individuals are found in the Parque Nacional de Los Haitises, on the northern slopes at the western end of the Cordillera Oriental. Fewer than a hundred pairs are thought to survive, making this one of the most threatened birds in the Caribbean.

Hispaniolan moist forests harbour endemic birds such as the grey-crowned palm tanager *Phaenicophilus poliocephalus*, the white-winged warbler *Xenoligea montana* and the colourful Hispaniolan trogon *Priotelus roseigaster,* recognized by its green upper parts, red belly and blue tail. The dark head, back and wings and orange belly of the La Selle thrush *Turdus swalesi* can be spotted in forests in the southwest of the country, such as those in the Sierra de Bahorucos National Park, also home to the Hispaniolan hutia *Plagiodontia aedium* and the giant tree frog *Osteopilus dominicensis*. The frog provides one of the ingredients for 'zombie powder' prepared by bokors or voodoo priests in Haiti.

Another secretive inhabitant is the Hispaniolan solenodon *Solenodon paradoxus*, which differs from its Cuban cousin in having its extremely mobile cartilaginous snout joined to the skull by means of a unique kind of ball-and-socket joint. It also emits streams of high-frequency clicks that are thought to be a simple and primitive form of echolocation.

The jewel in the Dominican Republic's crown, however, is found not on the island itself, but 130 km (80 miles) off the north coast. Banco de la Plata (Silver Bank, named after the large quantities of treasure lost here), Banco de la Navidad (Christmas Bank) and the outer parts of the Bay of Samaná are

Enriquillo is the only salt lake in the world to have crocodiles

the republic's Marine Mammal Sanctuary and the temporary winter home of a large proportion of the northwestern Atlantic population of humpback whales *Megaptera novaeangliae*. The 12–15-metre (40–50-foot) long whales are present from January to April, and they are here to give birth, raise their calves and mate. The waters are often filled with the sonorous but melancholy moans of the male whales, either staking a place in their huge underwater mating ground or 'lek', or attracting the attention of passing females. As many as 4000–6000 whales visit each year, the largest concentration of humpbacks on the planet.

Wildlife on Puerto Rico, the smallest and most easterly main island in the Great Antilles, has followed the same trend as most other Caribbean islands, suffering considerably from deforestation, with the result that rare and endangered species are restricted to small pockets of remaining virgin forest.

Nevertheless, the island and its satellites have a diversity of habitat types, from mangroves at the coast to dwarf mountain forests inland. At 1338 metres (4390 feet) above sea level, the highest peak is Cerro de Punta in the Togro Negro Forest of the Cordillera Central, home also to the 60-metre (200-foot) high Doña Juana Falls and to the green mango *Anthracothorax viridis*, a large hummingbird that feeds mainly on *Heliconia* flowers and insects, and is one of five species of hummingbirds native to the island.

Puerto Rico's karst country, like that of Cuba, has small, steep hills or *mogotes*, deep sinkholes and a labyrinth of underground passageways. Throughout the island, there are about 2000 cave systems, including some of the largest caves on the planet. At the Rio Camuy Cave Park, the river runs underground to form the world's third-largest subterranean river. Other notable formations include the 50-metre (170-foot) high cavern of Cueva Clara and the 120-metre (400-foot) deep Tres Pueblos sinkhole.

Karst country is the location of the last block of coastal dry forest and dry limestone forest on the island – the Guánica Dry Forest Reserve, a UNESCO Biosphere Reserve. The vegetation is adapted to months of dry weather, trees having thick trunks and waxy coatings to their leaves that keep in the water.

Here can be heard the haunting call of the Puerto Rican nightjar *Caprimulgis noctitherus*, a species known only from a few restricted areas in the dry southwest of the island. The reserve is also home to the rare yellow-shouldered blackbird *Agelaius xanthomus*, a bird that seems to be 'bullied' by everything else in the forest: the shiny cowbird *Molothrus bonariensis* behaves like the European cuckoo and lays its eggs in the blackbird's nest; the large swallow-like Caribbean martin *Progne dominicensis* takes

Puerto Rico's karst country has small steep hills or mogotes, deep sinkholes and a labyrinth of underground passageways

over its nest site; and the pearly-eyed thrasher *Margarops fuscatus* eats it eggs and kills its chicks. The blackbird is described by the IUCN as 'critically endangered', yet visitors sometimes spot it from hotels along the coast.

The western side of the island is the domain of just 150 sharp-shinned hawks *Accipiter striatus venator*. They can be seen hunting in the Maricao State Forest, where they pursue mainly small birds, which they catch on the wing.

In the northeast is Luquillo, a damp and dark tropical forest, where intense rainfall causes numerous landslides, and hurricanes account for localized damage, but which is an ideal habitat for the most numerous species of frog in Puerto Rico, *Eleutherodactylus coqui*. It is called the coqui frog because of its loud and distinctive call, described by local people as 'co-qui'. While most Caribbean frogs tend to be very small, this character is a whopper by Antillean standards, all of 40 mm (1½ inches) long for males and 50 mm (2 inches) for females. The coqui is also quite different from most other frogs in that the tadpoles develop inside the egg, so that fully formed froglets emerge.

The eggs are deposited not in a pond but in a cavity in a tree, such as a rolled palm frond petiole or sometimes the disused nest site of the banaquit *Coereba flaveola portoricensis*, Puerto Rican bullfinch *Loxigilla portoricensis* or Puerto Rican tody *Todus mexicanus*. The male guards the eggs meticulously, collecting moisture from anywhere with open water – bromeliads, ponds, ditches – to ensure that they are kept damp during dry spells and remaining on guard when the froglets hatch.

Adult coqui frogs spend their nights in the trees to avoid nocturnal ground-based predators, then just before dawn they clamber back down. One observer claims to have seen them parachuting by spreading their toes and body, but 'raining' frogs have yet to be proven.

Despite this predator-avoidance tactic, birds, mammals and the Puerto Rican racer snake *Alsophis portoricensis* catch older frogs, and youngsters are not immune from attack either. In their case, danger comes from a surprising quarter – the giant crab spiders *Olios* spp. Smaller froglets succumb easily, but larger individuals are able to escape the spiders' grasp by kicking out with their hind legs. The coqui itself is a voracious insect-eater. In some parts of the Luquillo Forest there are an estimated 20,000 frogs per hectare, and they consume 114,000 insects or spiders per hectare per night.

You might think that the population would have taken a tumble in 1989, when Hurricane Hugo slammed into the island and demolished much of the forest. Rather than reduce the frog's numbers, however, the subsequent increase in ground cover, with burgeoning growth of *Cecropia* and *Heliconia*, triggered a population explosion. Since then, it has been noticed that individual tree falls have the same effect, only on a smaller scale.

A more serious casualty of Hurricane Hugo was the bright-green Puerto Rican parrot *Amazona vittata*. Unlike the frogs, this seriously endangered species

The waters are often filled with the sonorous but melancholy moans of the male whales, either staking a place in their huge underwater mating ground or 'lek', or attracting the attention of passing females

ABOVE A humpback whale slaps its flukes on the sea's surface, a behaviour known as lobtailing – a form of communication.

RIGHT Silver Bank off the coast of the Dominican Republic is one of the main birthing sites for humpback whales in the western Atlantic.

was almost wiped out and did not immediately bounce back. It lives in the verdant 11,000 ha (28,000 acres) of the El Yunque Tropical Rainforest, known also as the Caribbean National Rainforest. The park has several different forest types, depending on altitude, with dwarf vegetation clinging to the steep slopes of El Toro – at 1076 metres (3532 feet) the park's highest peak.

The parrots are found in the rainforest on the lower slopes. They feed on the fruits, shoots and flowers of about 50 different plant species, especially the fruits of the sierra palm *Prestoria Montana* and tabonuco fruit *Dacryodes excelsa*. Natural predators include the pearly-eyed thrasher and the red-tailed hawk *Buteo jamaicensis*, the former also competing with the parrots for nest space.

At the time of Columbus, there were probably in excess of 100,000 Puerto Rican parrots, but by the 1950s there were just 200, and in 1975 there were 13. In 1989, the population began to increase slowly, but the hurricane hit them, and about 25 individuals survive today in a very small section of El Yunque, with a further 56 in captivity at the Luquillo Aviary. They were once found throughout the island, and, until 1989, there were significant populations on Vieques, Culebra Island and Mona Island.

ABOVE The male coquí frog, found in Puerto Rican rainforests, is especially noisy at night during the wet season of June to November.

OPPOSITE A banaquit feeds on a banana flower in Puerto Rico.

OPPOSITE The green iguana can grow up to two metres (7 feet) long, its tail making up over half its length.

Vieques, off the northeast coast of Puerto Rico, may not have its parrots any more, but it does have something else of interest – a phosphorescent bay. At night, the waters of Mosquito Bay are filled with bioluminescent marine organisms that glow brightly when disturbed. A dip in the bay is accompanied by an eerie blue-green glow, by which it is said to be bright enough to read a book. The location of a fish near the surface is clearly visible, its track lit up by the luminescent creatures. The organisms responsible are dinoflagellates *Pyrodinium bahamense*, microscopic algae that flash when agitated, a natural defence mechanism that startles predators. The narrow mouth of the bay prevents them being flushed out by the tide, and so millions of them live here (720,000 per gallon of water), with the tannins (rich in vitamin B12) from decaying mangrove roots and leaves providing the nutrients on which they thrive. Puerto Mosquito, as it is also known, is one of several so-called 'bio-bays' in the area, but it is by far the brightest.

Culebra Island has the Mexican bulldog or fishing bat *Noctilio leporinus*. It flies close to the surface and drags its clawed toes through the water to hook a fish. It is not exclusively a fish-eater, though: in the wet season, it feasts mainly on beetles and moths, which it catches in mid-air in the manner of other insectivorous bats. These fishing bats also hunt in the bioluminescent bays, the disturbance caused by their trailing claws clearly visible on the water surface.

Mona Island was once a stronghold for pirates and great riches are rumoured still to lie buried here. Today, it is known as the 'Galápagos of the Caribbean', a treasure house of wildlife with mangroves, coral reefs, pure white-sand beaches, 60-metre (200-foot) high sea cliffs and the world's largest system of sea caves.

The caves, which penetrate up to 250 metres (800 feet) into the limestone, contain indigenous pictographs, depicting bats and other cave dwellers, and some caverns were once sites of a thriving guano industry, exploiting bat droppings. The droppings of the local Mona iguana *Cyclura stejnegeri*, on the other hand, are utilized by terrestrial hermit crabs *Coenobita clypeatus*. They pick them apart and eat them. Like the crabs on Cuba, they emerge from their sheltered habitat on the plateau and head for the sea to spawn, in a mass migration occurring shortly after the August new moon.

Between April and September, Mona's beaches are invaded from the other direction – the sea. They host 160 nests of the hawksbill turtle *Eretmochelys imbricata*, the labours of 30–40 adults. Loggerhead *Caretta caretta*, green *Chelonia mydas* and leatherback *Dermochelys coriacea* turtles also come here to nest, peak activity occurring in August and September.

Another important reptile is the very rare endemic Mona boa *Epicrates monensis*. It grows to about 1 metre (over 3 feet) long, hiding in crevices by day and emerging to hunt at night. Young boas fall prey to the introduced rats, so the population is teetering on the brink of extinction ... but adult boas sometimes get their own back – they eat the rats!

3

LEEWARD ISLANDS

WHILE THE GREATER ANTILLES encompass the larger islands on the northern side of the Caribbean Sea, the Lesser Antilles include a long arc of smaller islands on the eastern and southern margins. The first chain to the east of Puerto Rico is known as the Leeward Islands, a name arising from sailing-ship days. They are the most northerly islands in the Lesser Antilles and therefore in the lee of the prevailing winds that blow from south to north. In geological terms, they are lined up along the boundary between two oceanic plates, at the point where the North American plate is disappearing below the neighbouring Caribbean plate; about 120 km (75 miles) north of the islands, the subduction zone, as this convergence of plates is known, is marked by the Puerto Rico Trench. Its lowest point, the Milwaukee Deep, is 8605 metres (28,232 feet) below sea level, the deepest submarine depression in the North Atlantic. At the bottom of this gash in the ocean floor the world's deepest-known fish, a 20-cm (8-inch) long cusk-eel or brotulid *Abyssobrotula galatheae* lives in the darkness 8372 metres (27,467 feet) down.

The closest of the Leeward Islands to Puerto Rico are the Virgin Islands, named Las Once Mil Vírgenes (11,000 Virgins) by Christopher Columbus on his second voyage, in an obscure reference to St Ursula and her maidens, who sacrificed their lives in Cologne. Once a hideout for pirates and buccaneers and now a hideaway for luxury yachts, the islands are today administered by the United Kingdom and USA.

The British Virgin Islands sit on an easterly extension of the Puerto Rico Bank and include Tortola, Virgin Gorda, Anegada and Jost Van Dyke (named after a Dutch pirate), plus 40 or so smaller uninhabited islands. All except Anegada, which is a flat, coral island, were uplifted from submerged volcanoes and are composed of volcanic debris and metamorphosed sediments. Most have fringing coral reefs and mangrove-lined lagoons. The original natural vegetation was cactus scrub and dry woodland, which includes the rare woolly nipple cactus *Mammillaria nivosa* and the Caribbean mayten tree *Maytenus cymosa*, two endangered plants these islands share with Puerto Rico. Today, all the islands have suffered degradation from human activities: we are responsible for the introduction of the principal land mammals – the rat and feral cat – but plenty of native birds and lizards, and countless fish remain.

Torola is the largest island in the group and has the most people, but the most significant for their wildlife are Virgin Gorda and Anegada. The former gets its name from its shape – mountains in the north and central part and lowland in the south. The Spanish word *gorda* means 'fat' or 'pregnant', and Columbus used it to describe the island when he first spotted it on the horizon. Since then it has been used as a base by both Bluebeard and Captain Kidd to harass passing Spanish shipping.

The highest point on the island is Gorda Peak, 418 metres (1370 feet) above sea level, but its most famous landmark is on the coast. Virgin Gorda Baths are a maze of weathered and eroded, house-sized granite boulders, together with crystal-clear pools and large caves, which were pushed up during a period of

PREVIOUS PAGES Aerial view of holiday islands, islets, beaches and reefs in the British Virgin Islands.

OPPOSITE The blue tang changes from a yellow juvenile (via a striped adolescent phase as shown in the picture) to an overall powder blue adult.

ABOVE The marlin was a leading character in Ernest Hemmingway's *The Old Man and the Sea*, set in the Caribbean.

uplifting and faulting about 15–25 million years ago. Endangered species unique to this area include a ground snake *Alsophis portoricensis anegadae* and a worm snake *Typlops richardi naugus*. The tiny gecko *Sphaerodactylus pathenopion*, also found here, was discovered in 1965 and once held the title of world's smallest lizard, an honour it now shares with the *Jaragua sphaero*, *Sphaerodactylus ariasae* from Beata, an island in the Dominican Republic.

Anegada is about 16 km (10 miles) long, 4 km (2 miles) wide, with its highest point just 8 metres (26 feet) above sea level. It has seemingly endless, isolated, white-sand beaches, those on the north coast hosting egg-laying hawksbill turtles *Eretmochelys imbricata*. The 30-km (18-mile) long Horseshoe Reef, part of a fringing reef that surrounds the island, is another of the Caribbean's giant stretches of coral. Huge electric-blue raiding parties of blue tang *Acanthurus coeruleus* engulf gardens of elkhorn, big brain and boulder corals, and in shallow waters slender, silvery bonefish *Albula* spp. feed on the bottom, their tails flapping above the surface.

Anegada is on the windward edge of the Puerto Rico Bank, where the sea floor beyond the Horseshoe Reef plunges from 55 metres (180 feet) to 365 metres (1200 feet). Here, huge feeding schools of Caribbean reef squid *Sepioteuthis sepioidea* communicate with 40 or more different body patterns. Pigment cells in their skin expand and contract with lightning rapidity so that their bodies ripple with stripes, chevrons and spots that say, 'Keep away' or 'I'm interested in you'. They are joined by surface-skimming flying fish of the family Exocoetidae, which can take off with a flick of the tail, and they are all pursued by yellowfin tuna *Thunnus albacares*, the staple fish of the tinned-tuna industry, and prized gamefish like the wahoo *Acanthocybium solandri*, dorado *Coryphaena hippurus* and the majestic blue marlin *Makaira nigricans*, one of the fastest fish in the sea.

Inhabited today by no more than a hundred people, Anegada has always been remote and sparsely populated, but before the age of enlightenment its wildlife was killed indiscriminately. The crews of passing ships once rounded up young flamingos on the island's extensive saltpans and herded them on board as food. The feral cats they introduced kill young Anegada rock iguanas *Cyclura pinguis* to this day. The island's importance as a wildlife site, however, has now been recognized. Caribbean flamingos *Phoenicopterus ruber ruber* were reintroduced in 1991 and are breeding successfully again, and a cat-neutering programme is beginning to help boost the lizard population.

The island may be flat and covered in scrub, but among the vegetation are several important plants, such as the gumbo limbo or turpentine tree *Bursera simaruba*, known locally as the 'tourist tree' because it has red and peeling bark. This species and the wild white-flowered frangipani *Plumeria alba* store moisture in their branches and trunk, an adaptation to the dry conditions. The endemic pokemeboy *Acacia anegadensis* is a dense, thorny acacia planted for shade.

Of the smaller islands in the British Virgin group, Guana is just 340 ha (850 acres) in area and 246 metres (806 feet) high, yet is considered to have one of the richest faunas of any island its size. It is privately owned and kept as a wildlife sanctuary – the only one in the world said to have a 'cocktail hour'. Flamingos and rock iguanas have been reintroduced, the 'guana on Guana' growing to a larger size than on other islands – up to 1.8 metres (6 feet) long. The island's salt pond has breeding pairs of black-necked stilts *Himantopus mexicanus*, Antillean pintails *Anas bahamensis* and Wilson's plover *Charadrius wilsonia*, and it is visited by little blue herons *Egretta caerulea*, yellow-crowned night herons *Nyctanassa violacea* and the rare masked booby *Sula dactylatra*. There are large colonies of brown boobies *Sula leucogaster* and brown pelicans *Pelecanus occidentalis*, the latter harassed by laughing gulls *Larus atricilla* that perch on the pelicans' heads hoping for fish scraps. Of the pigeons and doves, the bridled quail-dove *Geotrygon mystacea* is the local treasure. This species was brought to the edge of extinction through its reluctance to fly and its innate curiosity. Feral cats could catch it without effort and a man with a gun

The crews of passing ships once rounded up young flamingos on the island's extensive saltpans and herded them on board as food

could wipe out an entire population with ease. Now it is protected, and its numbers have recovered sufficiently to remove it from the critical list.

OPPOSITE The widely ranging black-necked stilt is found in estuarine, lacustrine, salt pond and emergent wetland habitats throughout the Caribbean.

Carrot Rock, off the southernmost tip of Peter Island, is no more than a small rock above water, yet it seems to have its own species of anole lizard *Anolis ernestwilliamsii* and skink *Mabuya macleani,* both of which are found nowhere else in the world. Norman Island, at the southern tip of the archipelago, is known locally as 'Treasure Island' because it is believed to have inspired Robert Louis Stevenson when he was writing his classic novel. Its coast is riddled with dark caves where corals can be seen feeding as if it were night-time. Near by, Pelican Island has the Indians, a series of rock pinnacles, around which swirls a host of reef fish, including a shoal of dark-loving glassy sweepers *Pempheris schomburgki* and commensal shrimps that seek the safety of lush underwater gardens of sun anemones. The island itself has Spy Glass Hill, once used by pirates to look out for treasure ships in what is known now as the Sir Francis Drake Channel, but was then Freebooters' Gangway. The salt ponds on the aptly named Salt Island once provided the salt ration for the Royal Navy, and during the dry-season salt harvest, when evening festivities preceded the official 'breaking of the pond', salt reapers had to give the government one bag of salt for every three harvested. Today, just one salt harvester retains the tradition.

ABOVE The leopard anolis *Anolis marmoratus* is endemic to Guadeloupe.

One of the world's most famous wreck dives is also found at Salt Island. The remains of the Royal Mail Steamer *Rhone,* which went down in a hurricane in 1867, lie on a reef 6–24 metres (20–80 feet) deep. The ship was heading for deep water, and the passengers were strapped into their seats for safety. When the storm hit, the captain was washed from the bridge and was never seen again. Cold seawater entered the engine room, enveloped the boiler, and it exploded, breaking the ship in two. It went to the bottom with only 23 of the 147-strong passengers and crew surviving.

LEFT The wreck of the RMS Rhone in the British Virgin Islands is one of the most sought-after dive sites in the Caribbean.

Made famous in the Hollywood feature *The Deep*, starring Jacqueline Bisset, the wreck is today encrusted with corals and sponges and awash with reef fishes. A lone great barracuda *Sphyraena barracuda* patrols its decks, and the purple egg patches of sergeant-major fish *Abudefduf saxatilis* decorate its cabin walls.

The US Virgin Islands consist of three main islands – St John, St Thomas and St Croix – and a host of smaller islets. St John and St Thomas are volcanic in origin, St Croix a raised coral reef. Of the three, St John is the least developed, but the most interesting from the ecological point of view. Over 50 per cent of its land area is a national park (which extends to include Hassel Island in St Thomas harbour) and designated a UNESCO Biosphere Reserve. Local variations in rainfall have produced several different forest types in a small area. Moist forests are found on the north shore and in the mountainous interior, with kapok *Ceiba*, mango *Mangifera*, sandbox *Hura*, saman *Samanea*, strangler fig *Ficus* and genip *Melicoccus* forming a canopy exceeding 25 metres (80 feet), and wild coffee *Psychotria* and teyer palm *Coccothrinax* the elements of a sub-canopy. Dry forests occur in the east and southeast, with dildo cactus *Cephalocereus nobilis*, prickly pear *Opuntia* spp. and Turk's cap cactus *Melocactus* spp. sharing space with shrubs such as maran *Croton* and thorn bushes like casha *Acacia farnesiana* and catch-and-keep *Acacia retusa*. Mangroves grow on the shore.

St Thomas is dominated by Drake's Seat, a famous viewpoint overlooking Drake's Passage, where the national flower, yellow cedar *Tecoma stans,* adorns the slopes. Yellow sets the theme here, for the national bird is the yellow breast *Coereba flaveola*. The island has a marine sanctuary, in which nurse sharks *Ginglymostoma cirratum* breed and courting humpback whales *Megaptera novaeangliae* visit between January and April.

St Croix has Buck Island Reef National Monument with an elkhorn-coral barrier reef surrounding two-thirds of the island. Much of it was devastated by Hurricane Hugo in 1989, and so the island, including its colonies of least terns *Sterna antillarum*, brown pelicans and frigatebirds *Fregata magnificens,* is now a study site for monitoring the recovery of wildlife and the environment after such a catastrophic event.

BELOW LEFT A spotted cleaner shrimp *Periclemenes* hides amongst the tentacles of a giant anemone despite them having dangerous stinging cells.

BELOW Breeding nurse sharks are a tourist attraction in St Thomas, and young sharks can be petted at the Coral World Ocean Park.

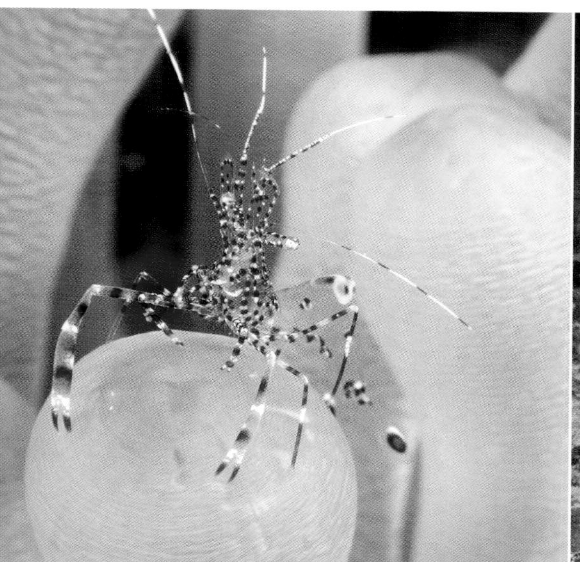

The most northerly of the Leewards, Anguilla is a small, flat coral island, shaped like the eel that gives it its name. It is about 26 km (16 miles) long and 3 km (2 miles) across at its widest, with its highest point a little over 70 metres (200 feet) above sea level. It is the quintessential island paradise, with beaches described as the best in the Caribbean and waters in six protected marine parks that range in colour from translucent turquoise to deep navy – 'tranquillity wrapped in blue', according to the local tourist office.

PREVIOUS PAGES Unspoilt beaches and forest in the British Virgin Islands.

ABOVE Newly hatched and camouflaged least tern chicks can be seen at any of 15 breeding colonies on St Croix.

About 80 per cent of the island is uncultivated and dominated by scrub and low trees, habitat favoured by the extremely rare Lesser Antilles iguana *Iguana delicatissima*, a shy and retiring species found in limited numbers on several Caribbean islands. About 20 brackish-water ponds are magnets for resident and migrant birds, and the mangroves are home to the mangrove cuckoo *Coccyzus minor*. Limestone and sandstone sea cliffs rise up to 30 metres (100 feet) above the sea on the northern side of the island, some of the sandy cliffs providing crevices and burrows for nesting red-billed tropicbirds *Phaethon aethereus*. Small bays have brown pelicans, royal terns *Sterna maxima*, turnstones *Arenaria interpres* and sanderlings *Calidris alba* feeding.

The remainder of Anguilla's rocky coast is limestone pavement, where cactuses – especially the pope's head cactus *Melocactus intortus* – grow. Small birds such as banaquits *Coereba flaveola*, grassquits *Tiaris* spp. and yellow warblers *Dendroica petechia* feed and nest here. On the offshore islands the colonies of breeding seabirds include tens of thousands of terns and noddies that arrive each April from South America.

In 1995, Anguilla had unexpected immigrants. Hurricane Luis, followed closely by Hurricane Marilyn, hit the region, and to the south of Anguilla Guadeloupe's forests were thrown into the sea. The floating vegetation drifted over 300 km (200 miles) northwards, and it carried passengers, in the form of green iguanas *Iguana iguana*. The animals spent over a month at sea marooned on their rafts, only to come ashore eventually on the beaches of Anguilla. It is thought that at least 15 individuals made landfall here, so now green iguanas live and breed on an island where they had not been seen before. This is probably one way in which many of the region's animals colonized the islands over the centuries.

At one time the Leeward Islands must have had a significant mammalian fauna, conspicuous today by its absence. On Anguilla, the cave known as Pitch Apple Hole contains fossils of mysterious creatures that are no longer with us, including an enormous hutia *Amblyrhiza inundata*, thought to have been the size of a black bear! These giant rodents lived about 125,000 years ago when the sea level was lower, and Anguilla, St Maarten, Saba and St Barts were a single island.

Anguilla was once linked politically with St Kitts, but in 1967 the locals rebelled, packing the St Kitts policemen back to their island. Paratroopers from Britain landed to restore order, an event that became known as the 'Bay of Piglets'. Today, it is a self-governing overseas territory of the United Kingdom.

To the south of Anguilla lies St Maarten/St-Martin. No more than 96 sq km (37 sq miles) in area, it is nevertheless divided politically, making it the smallest landmass in the world to be shared by two nations – the Netherlands in the south and France in the north. It has green, rolling hills, beaches, salt ponds and lagoons, and is host to an unusual range of residents from domestic goats and pigs to feral donkeys, African vervet monkeys introduced by the French 300 years ago, and a herd of cattle led by a very large black bull. The wildlife is more conventional, with iguanas, geckos, frigatebirds, pelicans and egrets.

Moving southwest from St Maarten/St Martin, the tiny island of Saba is another Dutch possession. It has no beaches, for it is a solitary, dormant volcano surrounded by sheer cliffs that rise steeply from the seabed to the peak at Mount Scenery, 887 metres (2910 feet) above sea level. One of Saba's claims to fame is the airport at Flat Point, considered to have the shortest commercial runway in the world. It is just 400 metres (1300 feet) long, of which only 300 metres

(1000 feet) are technically usable, with no overrun either at the ends or on the sides – just cliffs that plunge into the sea 60 metres (200 feet) below. According to pilots, take-offs and landings are similar to those to and from an aircraft carrier, except that the island cannot be pointed into the wind!

The rest of Saba is equally dramatic, with cactus and scrub on the lee side and thicker vegetation, including breaks of tree ferns and palms, cloaking the mountainous interior; closer to the peak, the cloud forest contains unexpectedly tall mountain mahogany *Freziera undulate*. Hidden on the green slopes are endemic lizards *Anolis sabanus*, and there are iguanas and racer snakes *Alsophis* spp. too. Tropicbirds nest on the sea cliffs, and bridled terns *Sterna anaethetus* and brown noddies *Anous stolidus* also breed on the island. Red-tail hawks *Buteo jamaicensis* and American kestrels *Falco sparverius* hover over the lower slopes, bridled quail-doves or wood hens are seen in the lower forests, while the cloud forests have the Saba trembler *Cinclocerthia ruficauda pavida* and purple-throated carib *Eulampis jugularis* and the green-throated carib *E. holosericeus*.

Diving off Saba is also exciting, for within a short distance of the land the seabed plummets down hundreds of metres, beyond diving range. Instead, divers head for pinnacles and seamounts that rise from the sea floor, including Eye of the Needle, Mont Michel and Shark Shoal (where ironically there are no sharks). Shark Shoal has two underwater peaks separated by a saddle, and the colour is breathtaking, for each is covered with red, yellow and purple sponges and corals, and kissed by silvery schools of snappers and horse-eyed jacks.

About 27 km (17 miles) southeast of Saba is St Eustatius, and at its southern end is the Quill, a huge volcanic cone the highest point of which is Mazinga Peak at 600 metres (1969 feet). The base of the dormant crater is 273 metres (896 feet) above sea level, and it contains a rainforest with ferns, bromeliads, orchids and banana plants. The Quill's protected area extends from an elevation of 250 metres (820 feet) down to sea level and beyond (for a marine park surrounds the island), and is host to red-bellied racer snakes *Alsophis rufiventris*, the latter growing up to 2 metres (6½ feet) in length, and two anole lizards *Anolis bimaculatus* and *A. wattsi*.

At the northern end of the island are the five hills of the Boven Subsector – the Boven itself, plus Venus, Gilboa, Signal and Bergje. The area has been badly disturbed by agriculture and domestic stock, but some vegetation and wildlife have survived, including such ubiquitous characters as the local ground lizard *Ameiva erythrocephala*, 15 species of orchids and the endemic statia morning glory *Ipomoea sphenophylla*, which is found only around Bergje Hill.

St-Barthélemy (or St Barts, as it is commonly known) is one of the French Antilles, yet its main town Gustavia is named after a Swedish king, the consequence of a century (1785–1878) when the island was a Swedish colony. Steep hills divide St Barts into several valleys that run to the sea, each with its own distinctive flora. Where wind and sea mist limit plant growth, the native

OPPOSITE The purple-throated carib is one of two hummingbirds to be found on the tiny island of Saba.

ABOVE Magnificent frigate birds roost and nest among the mangroves of Codrington Lagoon, Barbuda, a sanctuary accessible by boat only.

frangipani thrives. Known in patois as the *bois couleuvre* or snake tree, its flowers give off a scent like jasmine, and they are edible, an ingredient in a local jam. In dry valleys, the red-orange flowers of the geranium tree *Cordia sebestena* brighten the arid landscape. More sheltered valleys have the poui or trumpet tree *Tabebuia pallida* with its large mauve, rose or white flowers.

Like many (but not all) of the Leeward Islands, St Barts also provides a vital home for the Lesser Antillean iguana, a slightly smaller species than the common or green iguana with which it is sometimes in competition for living space and food. Adults are dark grey, but hatchlings are bright green, affording them protective camouflage from their natural enemies – snakes and raptors. This iguana lives in many habitats, including xeric scrub, dry scrub woodland, littoral woodland and mangrove, as well as lower-altitude rainforest. It is a herbivore, feeding in the morning on the leaves, flowers and fruits of the caper *Capparis spinosa,* Surinam cherry *Eugenia uniflora,* aquatic soda apple *Solanum tampicense* and white cedar

Tabebuia heterophylla. Its diet also includes manchineel, a plant containing toxic compounds that prevent other animals from eating it, but which fail to deter iguanas.

St Kitts of the Federation of St Kitts and Nevis, islands that attained independence from Britain in 1983, is known for its vervet monkeys *Cercopithecus aethiops* – not natives, but pets that were introduced from West Africa and kept by parasol-twirling French ladies in the eighteenth century. The monkeys escaped into the plantations and grounds of large estate houses, spread across the island and have since wreaked ecological havoc. Along with the introduced mongoose *Herpestes auropunctatus*, brought in to eradicate rats living among the sugar cane, the vervets wiped out all the ground birds, parrots and the St Kitts bullfinch *Loxigilla portoricensis grandis*. Today, they feast on sea grapes *Coccoloba uvifera* and clammy cherries *Cordia obliqua*, and dig limpets off the rocks on the shore, although the largest feral population is in the forested crater of dormant Mount Liamuiga (known formerly as Mount Misery). Here, the monkeys not only enjoy the high life but also get high. The bark and pods of a tree growing in the crater forest, known as the jumbie cutlass or coral tree *Erythrina corallodendron*, contain a narcotic alkaloid making it hallucinogenic.

The youngest of the island's four volcanoes, Liamuiga, grumbled away with a storm of 186 earthquakes as recently as 1988, causing several tourists climbing to the crater rim to hit the ground. It last erupted in 1843 – or so it is thought, for the records are unclear. A crater lake, lush rainforest and several sulphur vents serve as a gentle reminder of the island's volcanic origins.

The easternmost of the Leeward Islands is Barbuda, where Codrington Lagoon hosts over 5000 magnificent frigatebirds *Fregata magnificens*, the largest colony of its kind in the western hemisphere and the second largest in the world. It was recognized internationally in October 2005 when Antigua and Barbuda joined the Ramsar Convention, and Codrington was designated the island's first Wetland of International Importance. In October each year, activity here increases significantly when the conspicuous glossy-black males, each with his bright-red chest puffed out to its maximum size, vigorously court the females. The air is filled not only with the sounds of a galaxy of seabirds but also the drumming made by the male frigatebirds on their inflated sacs. Not long after, nests are constructed, but, on a relatively small desert island, sticks are at a premium and pilfering from neighbours is rife. At times like these, frigatebirds live up to their reputation as the real 'pirates of the Caribbean'.

The lagoon also harbours a lobster-breeding industry that sees its produce shipped to the best restaurants all over the world, although there is some concern that too many lobsters are being harvested, and conservation measures have been put in place. The island itself is made of coral, has pink sand beaches and is riddled with caves and sinkholes. Its brackish-water pools dry out to provide crystalline sea salt that is still harvested today. Curiously, the national animal is an import from Europe – the fallow deer *Cervus dama*, which was introduced to Barbuda and neighbouring Guiana Island by the lessees, the Codrington family, at some unknown date. By 1784, there were an estimated 3000 of them, and they were stripping the island bare. Today, their numbers have been considerably reduced, but there are two varieties surviving, one normal deer colour and the other almost black.

Barbuda's history is linked inextricably with that of neighbouring Antigua, the largest of the British Leeward Islands, about 50 km (30 miles) to the south. The eastern part of Antigua is underpinned by limestone; the central plain has marl and clay overlying volcanic material; the southwest is volcanic, with Boggy Peak in the Shekerley Mountains the highest point at 402 metres (1319 feet). This area of the island is the least overrun by people, but even here very little original vegetation survives, for 95 per cent of Antigua's forests were cut down to make way first for sugar plantations and then for tourist developments. Deforestation and introduced domestic goats, Indian mongooses and the ubiquitous brown rats *Rattus norvegicus* have all but destroyed native wildlife.

Casualties include the Antiguan racer *Alsophis antiguae*, a harmless, 1-metre (3-foot) long, green-brown snake that is heading rapidly towards extinction, along with the endemic Antiguan ground lizard *Ameiva griswoldi*. Fortunately, they survive on some of the numerous offshore islands and islets. The racer currently lives only on Great Bird Island, a plot of land no bigger than a city park, 3 km (2 miles) northeast of the main island; here the snake shares space with whistling ducks, pelicans and tropicbirds. In order to give its numbers a chance to recover, local conservationists have removed rats and mongooses from some of the islands, and racer snakes are gradually to be reintroduced to places they formerly occupied.

One place that is way out on a limb is the tiny, conical island of Redonda, located 55 km (35 miles) southwest of Antigua, between Montserrat and Nevis. It is 2.5 km (1½ miles) long by 800 metres (½ mile) wide and exactly 296 metres (971 feet) high, with precipitously steep cliffs that plunge into the sea. Thought once to have been used a rest stop for canoeists travelling among islands, it was known to the Caribs as Ocanamanru, but Columbus renamed it Santa María de la Redonda, meaning 'St Mary the Round'.

Interest in this remote lump of rock increased in the mid-nineteenth century when it was discovered that the huge population of roosting and nesting seabirds had contributed something that the rest of the world wanted – guano, in those days an important ingredient in the manufacture of explosives. Guano contains

OPPOSITE The Antiguan racer snake is one of the world's rarest snakes.

phosphates, used also as fertilizers in agriculture, and the phosphates were mined here. One of the horizontal shafts that were dug into the deposits, known as Centaur's Cave, can still be visited today. Mining stopped, but the equipment was maintained until 1929, when a hurricane blew away all the buildings.

In 1979, people went back – not mine workers but a group of naturalists, an archaeologist and a geologist. They discovered only two trees, for the tiny rocky island is covered mainly with coarse grasses and cactus. Nevertheless, they found a surprising diversity of animal life. Terns, boobies, pelicans and frigatebirds continued to contribute to the phosphate deposits, and the party also chanced upon a burrowing owl *Speotyto cunicularia*, which had been wiped out on Antigua by mongooses. The naturalists included an entomologist, so the tally of 20 species of small moth is not surprising, but the discovery of a beetle *Hymporus* spp. that is attracted to the smell of seabird guano was an unexpected bonus. Healthy domestic goats were also found, probably the descendants of stock left as emergency rations by pirates and buccaneers.

Today, the island is known less for its wildlife than for its whimsical royalty, particularly the famous 'King of Redonda', created by leading literary figures. It all started in 1865, when quarter-Irish Montserratian Matthew Dowdy Shiell annexed the island in order to give his only son a kingdom. On his fifteenth birthday Matthew Phipps Shiell thus became the first king of Redonda – King Felipe I. He moved to London and became a well-known writer. When he died in 1947, he was succeeded as king by British writer and poet John Gawsworth, one of the pseudonyms of Terence Ian Fytton Armstrong, who became Juan I. His reign was tempestuous, and for a fee many literary glitterati, including Lawrence Durrell, J. B. Priestley and Henry Miller, joined the Redondan aristocracy to fuel a life at the Alma Tavern in London. After Gawsworth's death in 1970, his literary executor John Wynne-Tyson became King Juan II. Wynne-Tyson abdicated, and the title passed to the Spanish writer Javier Marías – Xavier I – who created a literary prize named after the island. He too has been generous with Redondan titles: film director Francis Ford Coppola, for example, is the Duke of Megalópolis. Gawsworth's legacy, however, remains, and today many ennobled souls contest the title. It looks as if the tale of the King of Redonda will run and run.

Back on the main island, several of Antigua's geological features catch the eye, the most impressive being a row of sandstone pillars – the Pillars of Hercules – in the cliff side at the entrance to English Harbour. A second is Devil's Bridge, a natural limestone arch eroded by Atlantic breakers at Indian Town Point, the island's northeast extremity. It acquired the name 'Devil' because slaves escaping the sugar-cane estates were said to throw themselves into the churning waters there, and nobody ever came out of the sea alive.

The secluded waters of Windward Bay, near Falmouth Harbour, is one of few sites in the world where scientists dive in search of the exceedingly rare marine mollusc

Nutting's globe vase *Vasum globulus nuttingi*, a seashell closely related to a similar species on the other side of the Atlantic in West Africa. Groupers fail to see their conchological value and snaffle them at high tide.

At Snapper Point, conservationists are hoping to revive what was until the late 1980s one of Antigua's botanical jewels by excluding the island's feral goats. The site features four dramatic species of cactus that grow on a limestone pavement atop 34-metre (110-foot) high cliffs. Here the barrel-shaped Turk's head cactus and round pin-cushion species once grew prolifically alongside the more columnar dildo cactus and the so-called 'sucking cossie', a cactus with nasty hooked spines. Young ladies are said to have stuffed this cactus into old stockings and wielded it as a weapon against over-amorous males, for the spines are difficult to remove from the skin.

The human population, like the wildlife, has changed considerably over the centuries. First to arrive were the Ciboney people in 2400 BC, but by the time Columbus came here on his second voyage, in 1493, Arawak and Caribs were in place. Spanish and French settlements were taken over by the British in 1667, slaves were imported from Africa, and the changes continue. The current population of nearly 70,000 people on Antigua and Barbuda was increased dramatically almost overnight in 1995, when 3000 refugees settled here after fleeing Montserrat to the southwest.

Montserrat's *bête noire* is its 915-metre (3000-foot) Soufrière Hills volcano, an ongoing disaster area. It burst into life on 8 July 1995, causing the human population either to leave or to squeeze on to the opposite end of the island. And the eruption has not stopped. At the time of writing, in autumn 2006, glowing lava is erupting from the crater, a new dome is forming, and an ash and steam cloud extends nearly 100 km (60 miles) to the northeast.

All life on Montserrat has been affected. As on many Caribbean islands, endemic animals are clinging to life, and the last thing they need is to be

ABOVE All parts of the Philodenron species, seen here on the slopes of La Soufrière volcano, are poisonous because they contain calcium oxylate.

incinerated by a *nuée ardente* (glowing avalanche or pyroclastic flow). One of the casualties is the famous Montserrat mountain chicken *Leptodactylus fallax* – not a bird, but a large edible frog that gets its name from the culinary interest in its long back legs. At 20 cm (8 inches) long and weighing up to 700 g (1.5 lb), it is one of the largest frogs in the world; it can cover up to 2.5 metres (8 feet) in a single leap and lives usually in cool, moist forest above 300 metres (1000 feet).

The mountain chicken is noted not only for its gastronomic potential but also for its unusual breeding behaviour. Males attract females and deter rival males with a whooping call that carries up to a kilometre (over half a mile) through the forest. A female enters a burrow in the territory of a calling male, and here they mate. The eggs and tadpoles develop not in water but in a foam nest whipped up by the parents. The tadpoles are huge – up to 15 cm (6 inches) long, of which 12 cm (5 inches) is tail – and they live in the foam nest until they are froglets about 3 cm (1 ¼ inches) long. Then they leave and must fend for themselves in their especially hostile island world. They were once present on many of the islands of the Lesser Antilles, but today the only known survivors are on Montserrat, Dominica and in a captive-bred population at Jersey Zoo.

Other endemics threatened by habitat loss caused by the eruptions include: the Montserrat oriole *Icterus dominicensis oberi*, the tree bat *Ardops nichollsi montserratensis*, the Montserrat anole *Anolis lividus*, which is still common in mangroves, in acacia trees and on walls, and the Montserrat galliwasp *Diploglossus montisserrati*, a lizard known from one specimen collected in 1964, but rediscovered in 1998. Most of the devastation and the losses have been in the northern part of the island. The southern half remains relatively untouched by the volcano and is still very much the stunning 'emerald isle' that it always was.

That anything survives on Montserrat at all is something of a miracle. The island was hit by Hurricane Hugo in 1989 and in more recent years has taken a pounding from a succession of violent weather systems, including Tropical Storm Iris and hurricanes Luis and Marilyn. The land itself is very rugged, with two highland regions and many deep gorges. Most of the original forest was felled for timber, but secondary rain, cloud and moist forest and replanted trees covered about 70 per cent of the island before the eruptions. Now, much of the south part of the island has been flattened or burned.

The southern half of the island remains relatively untouched by the volcano and is still very much the 'emerald isle'

To the southeast of Montserrat is the French département of Guadeloupe, an archipelago with two main islands – Grande-Terre and Basse-Terre – which are shaped like a butterfly, its body marked by a narrow mangrove channel known as the Rivière Salée. There are also a host of smaller inhabited islands, including Marie-Galante, La Désirade, Iles des Saintes and Iles de la Petite-Terre.

Carbet Falls, the highest waterfall in the eastern Caribbean, is found in the Parc Naturel at the southern end of Basse-Terre. It has three cascades – the uppermost with a drop of 125 metres (387 feet), the second 110 metres (330 feet) and the lowest 20 metres (60 feet) – falling from the slopes of the island's active volcano, the 1467-metre (4813-foot) high Soufrière de Guadeloupe. With violent activity recorded as early as 1660, it was probably the first of the erupting Caribbean volcanoes to be observed by European settlers, and it last blew its top in 1977.

Geologically speaking, Guadeloupe lies at the southern end of the two distinct belts of the Leeward Islands arc and has characteristics of both. The eastern half – Grand-Terre – is composed of limestone from coral reefs resting on older volcanic rocks, a geology it shares with the Leewards' outer islands – Antigua, Barbuda, St-Barthélemy, St-Martin/St Maarten, Anguilla and Guadeloupe's own offshore islands Marie-Galante and Désirade – all of which face the Atlantic Ocean. The landscape here is of gently rolling hills and level plains. The western half – Basse-Terre – is dominated by more recent volcanoes and has a more rugged landscape with dense forest, a trait it shares with the inner islands – Montserrat, Nevis, St Kitts, St Eustatius and Saba, which face the Caribbean Sea.

To the west of these inner islands, isolated from the rest of the West Indies, is the tiny islet of Isla de Aves, named for its nesting seabirds. It is just 375 metres (1230 feet) long, 50 metres (164 feet) wide and 4 metres (13 feet) above sea level ... at least, on a calm day, for should a hurricane pass it is submerged completely! Birdlife can move away temporarily, returning when the sea has subsided, but the Venezuelan military, which has a scientific naval base here (much to the chagrin of other Caribbean islands, which contest Venezuela's right to claim an Exclusive Economic Zone), must stay through calm and storm. They can only do so because their tiny living quarters are built on stilts.

4
WINDWARD ISLANDS

ACTIVE VOLCANOES, boiling crater lakes, plunging waterfalls, black-sand beaches and mountains clad in lush 'oceanic rainforest' – these are the natural features that characterize the islands of the southern part of the Lesser Antilles, the Windward Islands. Like their northern neighbours, they get their name from the direction of the wind: they lay to windward of ships sailing from Europe, whose course across the Atlantic led them precisely to the dividing line between the Leeward and Windward island chains.

The closest Windward Island to this informal divide is the Commonwealth of Dominica (not to be confused with the Dominican Republic, see chapter 2), known locally as 'nature island'. With its dramatic white sea cliffs, sheer-sided valleys, spectacular waterfalls and deep river-filled gorges lined with tropical forest, it is probably as near as a visitor can get to the unspoilt wild Caribbean. It is often said that if Columbus returned to the region today, Dominica is probably the only island he would recognize, although its name when he was first here on Sunday, 3 November 1493, would have been Waitikubuli, from the Carib meaning 'tall is her body'. Many descendants of the Caribs (more accurately called the Kalinago people) still live on the island today, and every five years they elect a chief.

Dominica has exceptional natural resources. There are over 365 rivers – one for every day of the year – and the water in most is pure and drinkable. Of the many lakes, Boeri is the highest at 869 metres (2850 feet) above sea level, and Freshwater Lake the largest. There are many active volcanoes, and evidence of geothermal activity is everywhere. The famous Boiling Lake, surrounded by the steep cliffs of a large fumarole, is warmed by the heat from below, so its grey-blue water churns over at an average temperature of about 90 °C (194 °F) to produce an enveloping cloud of steam. It empties from time to time, as it did after an earthquake storm in 2004, but in 2005 it refilled and is currently about 60 metres (200 feet) across, its depth unknown; nevertheless, it is considered to be the second-largest lake of its kind in the world (after Rotorua on New Zealand's North Island) and a continuing reminder of the island's violent past and still active present.

Two small streams bring water from the surrounding hills to fill the Boiling Lake, and an outflow feeds the milky White River that cascades over the magnificent Victoria Falls. Near by is the Valley of Desolation, an area of steaming sulphur springs, clear-water pools, piping-hot geysers, bubbling, grey mud pots and multicoloured streams. Unlike the explosive volcanism of Montserrat and Martinique, Dominica is gently simmering on a low heat.

Even at the coast, sulphurous gases bubble away in Champagne Pool, a geothermal site at the end of a pebbly beach near the village of Pointe Michel; like everywhere on Dominica, volcanoes are never far away. Here, at the southwest tip of the island, Soufrière-Scotts Head Marine Reserve is a submerged volcanic crater, the sheer walls of which drop to unknown depths in a lava shoot, but its waters teem with marine life.

PREVIOUS PAGES
The waters of Diamond Falls in St Lucia drop 300 metres (1,000 feet) and leave the ground on the sides of the volcano at 40 °C (105 °F). They are thought to have curative properties.

ABOVE Steam rises from the floor of the Valley of Desolation in Dominica's Trois Pitons National Park.

Inland, the highest point is Morne Diablotin, an extinct volcano with sulphurous springs 1448 metres (4749 feet) above sea level. It gets its name from the rare diablotin or black-capped petrel *Pterodroma hasitata*, which was thought to be extinct but reappeared off the southeast coast of Dominica in 1984.

Another major peak, Morne Trois Pitons at 1402 metres (4600 feet), dominates the national park that shares its name. It has several crater lakes (including Boiling Lake) and many spectacular waterfalls, for which the island is rightly famous.

The waters of the beautiful Emerald Pool Falls reflect the surrounding rainforest, while Middleham Falls, with their clear, deep, plunge pool, are obscured by fine spray from their high cascade. In the east, the River Jack Waterfall, probably Dominica's tallest, can be seen from an accessible walking trail, while near La Plaine, the remote and plunging Sari Sari Falls are reached after a walk up a slippery riverbed. In the Roseau Valley area, the twin cascades of Trafalgar Falls have an upper fall fed by hot springs and a lower that is refreshingly cool; and Ti Tou Gorge, meaning 'little throat' in the local Creole patois, is a mysterious river-filled gorge with a small but perfectly hidden waterfall.

In the north of the island, the Brenner Falls are enclosed on three sides by vertical cliffs, giving the impression of dropping through a funnel; to visit the remote Brandy Falls a four-wheel-drive vehicle and tough footwear are needed.

One of the most unusual and private is Wavine Cyrique, where a 45-metre (150-foot) high cascade drops directly into the ocean at the Secret Beach. Then there are Spanny's Twin Falls, Flocky's Twin Falls, Dernière Falls, Coco Cascades, Space Mountain Falls and many, many more – all on an island with a land area of just 754 sq km (291 sq miles).

OPPOSITE The manicou or common opossum is consumed locally, its meat smoked then stewed.

In the northwest, the Northern Forest Reserve has some of the best-preserved oceanic rainforest in the region, and in the Syndicate Estate two of the world's rarest birds can still be found. The sisserou or imperial parrot *Amazona imperialis* is Dominica's national bird. One of the most striking of the Amazons, it has vivid purple feathers tipped in black on the head and breast, and deep-green feathers on the back, wings and tail. Male and female are alike, and mated pairs remain together for a life that can exceed 70 years. The wild population has been reduced to about 150–250 individuals, partly as a consequence of hurricanes damaging their high-elevation forest home. The smaller jaco, jacquot or red-necked parrot *Amazona arausiaca* lives on the lower slopes but is similarly threatened.

Of the other 176 recorded bird species, 59 are residents, and some are heard more than they are seen. On a trek to Boiling Lake walkers often enjoy the flute-like call of what is known locally as the mountain whistler – the rufous-throated solitaire *Myadestes genibarbis*. Hummingbirds are everywhere, and they are very visible. Dominica's forests are home to such elegant species as the Antillean crested hummingbird *Orthorhyncus cristatus*, purple-throated hummingbird *Eulampis jugularis* and colourful blue-headed hummingbird *Cyanophaia bicolor*, which lives only here and on Martinique. Hunting above the canopy may be the broad-winged hawk *Buteo platypterus*.

Two large mammals live on the island – the agouti *Dasyprocta anitillensis* and manicou (an opossum) *Didelphis marsupialis*. Both species were introduced from South America by ancient Amerindians many years ago but are now wild and widespread. The usual gamut of land-based reptiles is here too – iguanas, anole lizards (including the introduced Puerto Rico anole), ground lizards and four species of snake, including the boa constrictor.

Hidden away in the forest is the Dominican mountain chicken or crapaud *Leptodactylus fallax*. Like Montserrat's mountain chicken, this frog is threatened, but in this case the danger comes mainly from fungal and bacterial diseases that attack the skin and have spread through the population like wildfire. An estimated 70 per cent of the frogs have died, so the government is taking pro-active measures to deal with the tragedy. The frog is important

In the northwest, the Northern Forest Reserve has some of the best-preserved oceanic rainforest in the region

to the islanders, not only as a culinary speciality but also because it appears on the national coat of arms. Dominica's national tree is another rarity, the bois kwaib *Sabinea carinalis*, a shrub or small, spreading tree with red-to-crimson butterfly-like flowers. It occurs mainly on the west coast but is just one of an extraordinary number of plants growing throughout the island. It is thought that every major group of plant is represented here, including over a thousand species of flowering plants – orchids, bromeliads, vines, shrubs and a great diversity of trees. One of them, the buccaneer palm *Pseudophoenix sargentii*, with its characteristic bulging and constricted stem, was discovered only recently by botanists, although the locals knew it was here: they used the leaves for making hats. Today, it not as common throughout the region as it once was, and Dominica has not only the most southerly and easterly stand of these palms but also the highest above sea level, with trees growing at elevations up to 200 metres (650 feet).

OPPOSITE Knife-edge Ridge, Dominica.

Several forest types crop up in different parts of the island. Dry scrub woodland, with cactuses and agaves, is found in the west, while in the north and the south-central region rainforests occur between 300 and 550 metres (1000–1800 feet). Here the trees grow up to 34 metres (110 feet) tall, and dominant species include the gommier or candlewood *Dacryodes excelsa*, which exudes an inflammable white resin from wounds, the chataignier or breakaxe *Sloanea* spp., with extremely hard wood and massive buttress roots, and the carapite or amanoa *Amanoa caribaea*. A narrow transitional zone of 18-metre (60-foot) high montane forest is the place where the rasinier montagne *Podocarpus coriaceus* grows. Podocarps are an ancient group of conifers that were around when the supercontinent Gondwana broke up between 45 and 105 million years ago to create the main continents of the southern hemisphere; they are even found as fossils in Antarctica. This primeval forest leads upslope to elfin woodland of short, impenetrable trees that are buffeted by wind and rain, and enveloped in cloud at elevations above 900 metres (3000 feet). The dominant tree here is the gnarled and contorted kaklin *Clusia venosa*.

At sea level, swamp forests have the bwa mang *Pterocarpus officinalis*, anchored by huge buttress roots and growing to 27 metres (90 feet) tall, and there are both black and white mangroves along the sheltered west coast. The Atlantic east coast is buffeted by salt-laden winds, so the trees that grow here have thick and leathery leaves. Not many species do well under these conditions, but the survivors include white cedar *Tabebuia pallida*, galba *Calophyllum calaba*, sea grape *Coccoloba uvifera*, Indian almond *Terminalia catappa* and the zicaque or coco-plum *Chrysobalarrus icaco*.

Like many Caribbean islands, Dominica has land crabs that migrate in large numbers from forest to coast and back again each year to breed. Signs warn motorists that they are a protected species and not to squash them, although an open season for harvesting them supplies the main ingredient for 'crab backs', a traditional Creole dish of crab meat with chopped tomatoes, onions, garlic, breadcrumbs and seasoning, served in the shell or 'crab back'.

OPPOSITE Atlantic spotted dolphins are found in tropical and sub-tropical waters of the Atlantic.

From the direction of the sea come four species of sea turtle, which deposit their eggs on Dominica's fine sand beaches. Offshore, a regular procession of whales and dolphins might include short-finned pilot whales *Globicephala macrorhynchus* and false killer whales *Pseudorca crassidens*, spinner dolphins *Stenella longirostris* and Atlantic spotted dolphins *Stenella frontalis*, but the most impressive cetaceans in these waters are sperm whales *Physeter macrocephalus*. They come to the sheltered west coast to drop their calves and to mate, so mothers can be seen interacting with calves, and visiting bulls are sometimes spied courting females. They feed in the deep water close to the shore, where females dive down to 300 metres (1000 feet) and males to over 900 metres (3000 feet) in pursuit of squid and deep-sea fish and sharks. Several groups of whales visit the area each year; although from January to March a 'resident' group of seven individuals stakes a claim to Dominica's lee shore.

The whales are not confined to Dominican waters but range throughout the Windward Islands, including neighbouring Martinique, where 18 species have been seen or heard at one time or another. Similar conditions occur here, with sheltered, deeper waters close to shore on the west coast and more exposed, shallower waters to the east.

In the southwest, safe and sheltered anchorage once gave rise to a very unusual 'ship' – HMS *Diamond Rock*. It is actually a small volcanic island that lies 3 km (2 miles) offshore, with its peak 175 metres (574 feet) above sea level. In 1804, the British Navy ruled the waves, but ships were in short supply, and so to control this area Commodore Hood (later Vice-Admiral Sir Samuel Hood) had the bright idea of not deploying a ship to engage French ships entering the main harbours on Martinique, but commissioning an island. He armed it with 24- and 18-pounder cannons, created a field hospital in one of the caves and then caused havoc. The French Admiral Villeneuve was ordered by Napoleon to re-capture the rock that was causing his fleet so much grief, but the British 'crew' held out for months. The story goes that eventually barrels of rum were floated to the island, enabling the French to overcome the drunken British sailors and win the day.

Martinique, however, is probably best known not for the sound of its cannons but for explosions of a very different kind – those coming from Mount Pelée, a volcano at the northern end of the island. In 1902, the then capital of the island, St-Pierre, known as the Paris of the West Indies, was the French Caribbean's premier city, but after events in the spring of that year, it became instead a dead ringer for Pompeii.

On the morning of 8 May, Pelée suddenly burst into life. A deadly *nuée ardente* – a glowing cloud of searing hot gases, ash, pumice and glass shards – raced down a valley on the mountain's slope and, before anybody had a chance to escape, it engulfed St Pierre. It was precisely eight minutes to eight, the time the military hospital clock ceased to work and the telegraph to Fort-de-Franc went dead.

Just three people survived – a man and a small girl who both lived at the edge of town and a drunkard holed up in the underground gaol, who drank water running down his cell walls until he was rescued by a neighbouring parish's priest. Not all the residents perished, however. More than a thousand had heeded warnings prior to the eruption and fled to safety at the far end of the island; as for the rest, over 28,000 people were incinerated in a few seconds.

The volcano even produced its own monument to the dead. A gigantic pillar of lava, which became known as the Tower of Pelée, was pushed up into the sky. It was 300 metres (1000 feet) tall and 150 metres (492 feet) wide at its base. A year after the eruption it collapsed, but the volcano was not yet dead. It exploded again on 16 August 1929, flinging lava blocks from the 1902 dome into the air and sending glowing avalanches rolling down to the sea. Throughout that year it continued to explode with increasing ferocity. Four major eruptions occurred during September and October, and then the mountain settled down for a series of less violent outbursts until December 1932. At the time of writing, the 1397-metre (4582-foot) high mountain is dormant, but for how long? Pelée is still considered to be one of the deadliest volcanoes on Earth.

About 12 km (20 miles) to the south of Martinique are the remains of a giant volcano that once rose to over 3500 metres (12,000 feet), making it one of the highest peaks in the Americas, but all that is left today are two conical-shaped hills known as the Pitons, about 600 metres (2000 feet) above sea level. They are the most-recognizable natural feature on St Lucia, an island that also boasts the world's only 'drive-in' volcano, the dormant but steaming and bubbling Sulphur Springs. This entire area was once a gigantic strato-volcano, its centre marked by the sulphur fumaroles and hot springs to the southeast of the town of Soufrière. The Pitons – Gros Piton and Petit Piton, joined by the high Piton Mitan ridge – are the degraded cores of two lava-dome volcanoes that were like warts on the side of the collapsed giant.

The Pitons today are a UNESCO World Heritage Site, noted not only for their geology, but also for the unique vegetation on their slopes. Two small, endemic shrubs *Acalypha elizabethae* and *Bernardia laurentii* and a tree, the pencil cedar *Juniperus barbadensis* var. *barbadensis*, are found on the summit of Petit Piton and nowhere else. The variety of plants in such a small area is surprising. The steepness of the slopes has made them unsuitable for farming and means that small tracts of virgin rainforest remain, with at least 148 plants identified on Gros Piton and 97 on Petit Piton. This lushness and diversity in the flora has influenced the fauna too, with a fine collection of endemic animals that live or visit here. As many as 27 bird species have been recorded on Gros Piton, including five endemics: the St Lucia oriole *Icterus laudabilis*, St Lucia black finch *Melanospiza richardsoni*, St Lucia flycatcher *Myiarchus oberi sanctae luceae*, St Lucia peewee *Contopus oberi* and St Lucia house wren *Troglodytes aedon sanctae luceae*. The Pitons are an extraordinarily special place.

Other parts of the island are no less interesting. While the northern coastal areas of St Lucia or 'Helen of the West', as the island is sometimes known, are the realm of the tourist, the central and the southern parts have been less developed. The interior is mountainous, the highest point being Mount Gimie at 950 metres (3117 feet) above sea level, and there are streams and waterfalls everywhere. One of the most accessible is Diamond Falls, the lowest of six falls that cascade mineral-rich

water down the steep slopes of the extinct volcano. Louis XVI of France had a bathhouse built for his soldiers here, which has recently been restored.

ABOVE The St Lucia Parrot is found in the mountain rainforests of St Lucia, where it is the island's national bird.

Large tracts of forest have been preserved, including 7689 ha (19,000 acres) of National Rainforest, full of indigenous trees, giant ferns, tiny bromeliads and wild orchids. It is the natural habitat of the jacquot or St Lucia parrot *Amazona versicolor*, this island's only parrot. Predominantly green, with a cobalt head, scarlet breast and mottled, purple belly, it is conspicuous in sunlight, but in its dark rainforest home it blends in with its background so well that its raucous cackling is the only thing that gives it away. It was once common throughout St Lucia, but in 1902 the naturalist Lady Thompson wrote, 'Unfortunately dead birds may be found almost every week in the market of the little town of Soufrière, and they are eaten as a delicacy by both black and white Creoles.'

St Lucia's coastal forests and dry shrub areas have been home to one of the most deadly snakes in the Americas

ABOVE Dense rainforest on the holiday island of St Lucia.

RIGHT St Lucia's Soufrière Bay with the twin volcanic peaks of The Pitons.

Hunting for the pot was followed by habitat destruction and poaching for the pet trade, so that, by 1976, the population had dropped to just 150. In the nick of time, conservation measures were introduced, and the St Lucia parrot was adopted formally as the island's national bird. Today, the wild population stands at about 350–500, still small enough for the bird to be considered vulnerable but thought to be increasing – a conservation success story.

There may be a less happy ending to the story of the island's only poisonous snake. St Lucia's coastal forests and dry shrub areas have been home to one of the most deadly snakes in the Americas – the St Lucia lancehead or fer-de-lance *Bothrops caribbaeus*. It is a member of the pit viper group and has special sensory pits on its face that enable it to detect the heat from warm-blooded mammals. Its strike is extraordinarily fast, and on average it injects 105 mg (1⁄250 oz) of venom for each bite; half that dose is fatal for humans. The chances of meeting one are small, however, for few have been seen on the island in recent years and anyway, like most snakes, the fer-de-lance will shy away from human disturbance.

Local people will probably alert you to the danger nonetheless, but given the superstition that is rife on the island you are just as likely to be warned about such mysterious creatures as Papa Bois, the father of the forest, who takes his revenge on woodcutters for damaging his domain; the *diablesse*, a white demoness, who meets her victims on bridges and feeds them lizards and snakes; or the *soucouyant*, a vampire, who sheds her skin and flies across the sky. This last being can be destroyed only if you find her desiccated skin, dust it liberally with salt and pepper so that it is too itchy to put back on and then sprinkle her with holy water.

In the real world, St Lucia has three other snake species – the large boa constrictor *Constrictor constrictor*, also known as the *tête-chien* on account of the dog-like shape of its head; the tiny worm snake *Leptotyphlops bilineatus*, one of the smallest snakes in the world; and the St Lucia racer or couresse *Leimadophis ornatus,* which survives only on the Maria Islands, a nature reserve at the southern tip of St Lucia, although it has not been seen for some time.

Further south again, by about 15 km (25 miles), is the next of the Windwards – St Vincent, also an active volcanic island. Its volcano – another with the name La Soufrière – blew its top in 1718, 1812, 1902, 1971–2 and 1979, and they have all been violent explosions. The 1902 eruption, for example, which began on 6 May and ended on 30 March 1903, killed 1565 people. Today, the summit is often shrouded in mist, but steam is still emitted from the floor of the crater.

Wildlife here includes the endemic whistling warbler *Catharopeza bishopi* and the rare St Vincent parrot *Amazona guildingii*. At one time these parrots could be found from the mountains down to the sea, but by 1970 they were considered endangered, and in the 1980s the population hit an all-time low. The problem is not just their good looks but also their rarity, and the rarer they get the more expensive they become. On the illegal collectors' market a single specimen could command over $10,000 and, with demand outstripping supply in captive-bred birds, poaching is rife. Today, they are restricted to small patches of remnant rainforest in the mountains, but with protection and captive breeding their numbers have increased. Currently, there are about 500 St Vincent parrots in the wild, and it has become the island's national bird.

La Soufrière and parrots aside, St Vincent is probably most famous for an introduced plant. It was here that Captain Bligh brought breadfruit from Tahiti after the 'Mutiny on the *Bounty*' incident. At the time, breadfruit was rumoured to be a 'superfood' consumed by the superstrong Pacific islanders, and the British authorities reckoned that feeding it to plantation slaves would bolster their ability to work. In January 1793, the plants were placed in the botanical gardens at the capital, Kingstown, but the venture was to no avail, as the slaves refused to eat the alien food. The gardens, however, also have a specimen of the Soufrière tree *Spachea perforata*, which displays the island's national flower, but has not been seen in the wild since 1812.

St Vincent is not politically isolated, but is one in a string of islands, islets and cays, known as the Grenadines, that stretch southwards and are all part of the single nation St Vincent and the Grenadines. The islands were known to the early Spanish explorers as Los Pájaros, meaning 'the birds', because their silhouette on the horizon against the setting sun resembled a flock of seabirds in fight. Like the other Windward Islands, they have eastern shores pounded by the Atlantic Ocean, and sheltered bays and beaches on the western, Caribbean Sea side.

Next in line after St Vincent is Bequia. It is described as a 'sleepy island where time stands still' but waiting here is an unexpected time bomb – not a volcano, but a global concern, for Bequia is centre stage in the whaling debate. After the International Whaling Commission's Moratorium on Whaling in 1986, the island was given dispensation, under the terms of the aboriginal whaling clause, to kill two humpbacks per year. Bequia's whaling fleet consisted of one man and an open boat, but the man, one Athneal Olivierre, died in 2006. A close associate has been groomed to take over the business.

In addition to the annual humpback catch, pilot whales and other small cetaceans were also taken regularly by the crews of boats sailing out of Barrouallie on the leeward side of St Vincent. In 2000, they added an illegally caught Bryde's whale *Balaenoptera edeni* to their tally, a catch that coincided with Japan's proposal to catch Bryde's and sperm whales as part of its 'scientific whaling' programme. So, this 'sleepy' island of Bequia and her neighbours are right in the thick of an international controversy.

The Grenadines were known to the early Spanish explorers as Los Pájaros, meaning 'the birds', because their silhouette on the horizon resembled a flock of seabirds in flight

Next to Bequia is Mustique, the island of Basil's Bar and international celebrities. It is privately owned by 55 homeowners and run by the Mustique Company in collaboration with the St Vincent government. The island once had seven thriving sugar plantations, but the introduction of sugar beet elsewhere in the world put paid to that, and the entire island fell into disrepair. In 1958, it was bought by the Hon. Colin Tennant (Lord Glenconner), and in the following years he transformed it into an idyllic but very exclusive holiday haven.

The island itself is 567 ha (1400 acres) of rolling hills, deserted white-sand beaches and no traffic lights, beach vendors or litter; the speed limit is 32 km/h (20 mph). Bird-watching is best at the lagoon, where 50 indigenous Caribbean species are found, including the mangrove cuckoo *Coccyzus minor* and tropical kingbird *Tyrannus melancholicus*

With no courses on Mustique, the dedicated golfer must take the boat to Canouan, but there is more here than greens and bunkers (even if the course is designed by the legendary Jim Fazio and home to the Trump International Golf Club), for Canouan has one of the largest coral reefs in the Caribbean. The nearby islets of Tobago Cays, however, are considered the real heart of the Grenadines. Here, a 4-km (2½-mile) long horseshoe-shaped coral reef protects the islets of Petit Rameau with its mangroves, Petit Bateau, Baradal and Petit Tabac; the reef's windward edge is in turn protected from the Atlantic swells by World's End Reef. The cays are uninhabited, each with no more than a tuft of palm trees surrounded by white sand beaches, and the whole area, covering 50 sq km (19 sq miles), is protected by the Tobago Keys Marine Park – one of the locations, incidentally, of the desert-island scenes in the film *Pirates of the Caribbean*.

To the southwest of the cays are Mayereau, location of an ecologically important salt pond and also part of the Tobago Cays Marine Reserve, and the larger island of Union. With its high peaks, Union has been dubbed the Tahiti of the West Indies, but it came to prominence recently for purely herpetological reasons. In May 2005, amateur naturalist Father Mark de Silva discovered a spectacular little lizard here. Resembling tiny clusters of jewels, only brightly coloured males, just 30 mm (1¼ inches) long, have been found among rotting logs in the dry forests on the northern slopes of Mount Taboi. Co-describers from Avila University and Milwaukee Public Museum named it *Gonatodes daudini* (another dwarf gecko), in honour of Union Island conservationist Jacques Daudin.

The last of the Grenadines administered by St Vincent is its namesake Petit St Vincent, known locally as PSV – another unspoilt island paradise. The singer Cliff Richard has a house here, which in 2005 was visited by the British Prime Minister Tony Blair.

The remaining Grenadines – Petite Martinique, Carriacou and Ronde – are administered by Grenada, which lies a further 8 km (5 miles) to the south of Ronde. Unlike its neighbours, Petite Martinique is not predominantly a tourist island;

OPPOSITE False killer whales frequent the water of the Windward Islands and have been killed illegally by local whalers.

rather, it has been described as 'a place where island people return after a life at sea'. It is little more than the tip of a volcanic cone pushed above the sea's surface, but it is a pretty cone with fine sand beaches and natural harbours. Carriacou is relatively untouched too. Wooden schooners are still built here. It is hilly rather than mountainous, its highest point being High Point North, 291 metres (955 feet) above sea level. In its forests and scrub are manicous, red-legged tortoises *Geochelone carbonaria*, iguanas and boa constrictors, along with pelicans, terns, boobies and frigatebirds. Hawksbill *Eretmochelys imbricata*, leatherback *Dermochelys coriacea* and occasionally loggerhead *Caretta caretta* turtles nest on the beaches at the northern end of the island.

ABOVE The slopes of Grenada's extinct Grand Etang volcano are home to the nine-banded armadillo.

Ronde is a small islet between Carriacou and Grenada, and about 8 km (5 miles) to its west is an active underwater volcano with the forceful name of Kick-'em-Jenny. It rises about 1300 metres (4300 feet) above the sea floor but falls short of the surface by about 180 metres (600 feet). It was first noticed on

23 July 1939, when it broke the surface briefly during an eruption that saw steam and volcanic debris being ejected 275 metres (900 feet) into the air. The explosions caused a series of tsunamis, each about 2 metres (6½ feet) high, which reached the shores of the other islands in the southern Grenadines. Since then it has erupted 12 times, the last being in December 2001, and is the only Caribbean volcano likely to explode without warning. At the time of writing, there is a Yellow Alert notice in force with a 1.5-km (1-mile) exclusion zone around the volcano. It is thought it received its name from the fact that the waters in this area can be very rough, and was confused on maps with the nearby Ile Diamante or Diamond Island (not to be confused with Diamond Rock).

Grenada itself is known as the Spice Island, for many different spices, principally nutmeg, cloves, ginger and cinnamon, grow here. It once had its own active volcanoes, but in recent years the biggest impact on island life has been rogue hurricanes. Grenada was thought to be too far south to be much troubled by hurricanes, but that changed in 2004, when Hurricane Ivan, with winds gusting to 230 km/h (145 mph), slammed in, followed in 2005 by Hurricane Emily, both causing extensive damage on the main island and on Carriacou. The islands are recovering rapidly and life – both human life and wildlife – is getting back to normal.

Grenada's mountainous interior is topped by Mount St Catherine, 840 metres (2756 feet) above sea level. To the south, with peaks over 600 metres (2000 feet), is the Grand Etang National Park and Forest Reserve. It is crossed by small rivers with spectacular waterfalls, such as Concord, Au Coin and Fontainebleau, all of which flow from the highlands into the sea.

The extinct crater of Grand Etang mountain is filled with azure-coloured waters, the surface of the lake situated 536 metres (1760 feet) above the sea. Towering mahogany and gommier trees clothe the slopes and, apart from feral goats, donkeys and cattle, the large animals most likely to be encountered are iguanas, tree boas, manicous, nine-banded armadillos *Dasypus novemcinctus* and introduced mongooses *Herpestes auropunctatus*. The oddest addition is a troop of mona monkeys *Cercopithecus mona*, brought from Africa

Apart from feral goats, the large animals most likely to be encountered are iguanas, tree boas and nine-banded armadillos

300 years ago. Today, they are acknowledged kleptomaniacs that delight in stealing sunglasses and cameras from visitors.

About 150 bird species can be found throughout the island, highlights being the Grenada dove *Leptotila wellsi* and the hook-billed kite *Condrohierax uncinatus*. The dove survives in just two localities, with about 80 individuals in the Mount Hartman National Park, a dry cactus-scrub ecosystem in the south, and 20 at Perseverance. However, these figures date from before the hurricanes; few dove sightings have been reported since, and it is considered to be one of the most endangered birds in the world. The kite, on the other hand, is widespread in the Caribbean. Predominantly a snail eater, feeding on small arboreal snails, it will also take frogs, caterpillars or other insects if they are readily available. Another endemic that might be seen here is one this island shares with St Vincent – the Grenada flycatcher *Myiarchus nugator*.

Another wildlife haven is Lake Antoine, about 13 km (8 miles) north of the town of Grenville on the east coast. It is a shallow crater lake that has sunk to just 6 metres (20 feet) above sea level and is visited by several species of local birds, including the limpkin *Aramus guarauna*, a relative of the cranes and rails that is an avid devourer of apple snails *Pomacea* spp. Near the lake are the River Sallee boiling springs, which contain both clear and sulphurous waters.

At the coast, a number of sites are of interest. In the north, several beaches play host to sea turtles, with Bathways Beach in the Levera National Park a principal nesting ground. Levera also has mangroves and a lagoon – Levera Pond – that are important for black-necked stilts *Himantopus mexicanus*, common snipe *Gallinago gallinago*, herons and a host of waterfowl. In the south, the estuary at La Sagesse Nature Centre has mangroves and beaches with palm trees, but its primary magnet for birds is a salt pond. The brown-crested flycatcher *Myiarchus tyrannulus*, Caribbean coot *Fulica caribaea*, green-backed heron *Butorides striatus*, little blue heron *Egretta caerulea* and the northern jacana *Jacana spinosa* often appear.

Columbus dubbed the island Concepción in 1498, but later Spanish settlers thought it so resembled the green hills of Andalusia that they renamed it Granada, after the famous city there. The French revised this to Grenade and the British to Grenada. But although the island was easy to name it was difficult to settle. The local Caribs repelled all invaders for over a century and a half, but were eventually overcome, after a series of losing battles, by the French. Rather than submit to French rule the Caribs committed suicide, jumping to their deaths from high cliffs in the north of the island – a site that the French called Le Morne des Sauteurs, meaning Leapers' Hill.

OPPOSITE The little blue heron uses its sharkbill to stab at prey, but is only successful at catching anything 60 per cent of the time.

5

SOUTHERN CARIBBEAN

BARBADOS MAY be part of the Lesser Antilles but it is out on a limb, set apart from the rest of the Windwards, about 160 km (100 miles) to the east of St Vincent and the Antilles volcanic arc; it is the most easterly of the Caribbean islands. Although grouped with the Lesser Antilles, it is geologically and biologically quite different, with fewer mountains and less animal life. In bygone days it was also hard to reach. Sailing ships had difficulty in landing, because of the prevailing wind direction and position of the few natural harbours, making the island at first of little interest to the European settlers. Barbados remained British from the seventeenth century until 1966, when it gained its independence, so it has been more influenced by British culture than any of the other islands, gaining nicknames like 'Bimshire' and 'Little England'.

Geologically speaking it is a coral island, with a 90-metre (300-foot) thick layer of chalky ex-coral deposits sitting on shales, clays and conglomerates that have been uplifted from the ocean floor. The highest point is Mount Hillaby, which rises to 340 metres (1115 feet) in the north-central part of this triangular island. To

PREVIOUS PAGES A flock of vividly coloured scarlet ibis stand out from the blue of the sky and the green of the forest.

BELOW Castara Bay, on Tobago's north coast, has a crescent of coarse yellow sand, and is home to a 21-metre (70-foot) high almond tree estimated to be 150 years old.

the west the land drops in a series of green, rolling terraces, and in the east it meets the sea with high cliffs.

Surrounding most of Barbados is a series of coral reefs, including the North-east Coast Marine Park, where researchers are revealing how reef fish spend their day. They have discovered that these fish are very attached to their homes, few of them emigrating between reefs (the exception being jacks *Caranx latus* and *C. ruber*). Within their reefs, however, many species, including surgeonfish *Acanthurus bahianus* and *A. coeruleus*, filefish *Cantherhines pullus*, butterflyfish *Chaetodon striatus*, angelfish *Holocanthus tricolor* and parrotfish *Sparisoma viride*, range widely among the coral heads. The research has implications for fishing practices on all the coral reefs in the Caribbean, for once an area is fished out, this immobility means the area will repopulate very slowly.

The coral rock underlying most of Barbados (the exception being the Scottish Highland area where the coral has been eroded) is riddled with caves and caverns, many surviving as roosting and breeding sites for the island's eight species of bats.

In the centre is Harrison's Cave, which was mentioned in documents dated 1795 but disappeared from public gaze until it was 'rediscovered' in 1976. In 1981 it was opened to the public, so visitors can now view its wealth of stalactites and stalagmites and its lake and fast-flowing subterranean streams and waterfalls from the comfort of a miniature tram that runs through the passageways. It travels through a large cavern, known as the Great Hall, which is 30 metres (100 feet) high and 45 metres (150 feet) across, and stops in another that has thin, fragile, tube-like stalactites, known as soda straws, in its ceiling and two streams that cascade from a height of 14 metres (45 feet) at Twin Falls into a lake with green water. Next door, in the midst of a patch of jungle accessed via Jack-in-the-Box Gully, is Coles Cave. It has smaller passages that lead down to water, and is thought to be linked to Harrison's Cave, but the connection has yet to be found.

Some caves open to the sea. Animal Flower Cave at the island's northernmost tip is lined with yellow and purple sea anemones – the 'animal flowers' that give the place its name. Visitors do not have to dive, however, for the cave is currently above sea level, the result of Barbados rising at a rate of 25 mm (1 inch) every thousand years.

Across the island, very little original vegetation remains, the land having been stripped to make way for sugar plantations. In its place are introductions from all over the world – mahogany *Swietenia* spp. from Honduras, tamarind *Leucaena glauca* from Indonesia, casuarinas or mile trees *Casuarina equisetifolia* from Australia and the flamboyant or royal poinciana *Delonix regia*, a native of Madagascar, considered by some to be the most colourful tree in the world. These trees grow today alongside such natives as the Bajan ebony, shak-shak or mother's tongue *Albizia lebbek* and the manchineel *Hippomane mancinella*, which has poisonous apple-like berries, said in local folklore to be the 'apples' of the Garden of Eden. Whitewood trees *Bucida buceras* are often found in the vicinity of manchineel, the whitewood bark an antidote to the manchineel's poison.

In the north-central area are Turners Hall Woods, the last remnant of the tropical rainforest that once covered the island. Here, sand box *Hura crepitans*, trumpet tree *Tabebuia aurea*, the indigenous macaw palm *Aiphanes minima* and silk cotton *Ceiba pentandra* – a tree that, according to Bajan legend, walks at night – are still to be found. On the west coast, a patch of forest also survives at Joe's River. Bordered by the 300-metre (1000-foot) high Hackleton's Cliff on one side and the Atlantic Ocean on the other, it has mahogany, whitewood, cabbage palms *Sabal palmetto* and bearded fig trees *Ficus citrifolia*. The fig trees have long aerial roots, which gave the island its name: in 1536, the Portuguese explorer Pedro Campos saw them and called the island Los Barbados, meaning 'the Bearded Ones'.

Introduced animals such as European hares, Burmese mongooses and African monkeys characterize the large mammals, and bird life is restricted to a hundred species of transients and migrants and 20 residents, mostly doves, herons, egrets,

LEFT Silky or pygmy anteaters are usually hard to spot as they live high in rainforest trees, often sleeping by day.

hummingbirds and finches. There are whistling frogs and toads and a particularly noisy tree frog that keeps visitors awake at night. Three lizard species, the ubiquitous red-legged tortoise *Geochelone carbonaria* and visiting sea turtles represent the reptiles.

Protected areas are few on Barbados. In the southwest, Graeme Hall Swamp and Nature Sanctuary is the largest expanse of open water. It is fringed with mangroves, a roosting and nesting place for both resident and migrant birds. On the east coast, the Folkestone Marine Park features an artificial reef, the remains of the *Stavronikita*, a ship that caught fire in 1976 and was deliberately sunk. It lies in 36 metres (120 feet) of water, just 800 metres (½ mile) from shore, and is now home to corals, sponges and a host of reef fishes.

Trinidad and Tobago, on the other hand, have ten protected areas between them. These islands, about 130 km (80 miles) south of the Grenadines, are known affectionately as T & T, and the fact that they are just 11 km (7 miles) from Venezuela's northeast coast means the flora and fauna have a close affinity to their mainland neighbour, giving them the greatest diversity in wildlife of all the Caribbean islands.

There are over one hundred species of mammals alone, many of them familiar South American creatures such as ocelot *Leopardus pardalis*, collared peccary *Tayassu tajacu* and silky anteater *Cyclopes didactylus*, although bats account for the greatest number of species, including the notorious vampire bat *Diaemus youngi* that licks blood from the wounds it inflicts with its sharp front teeth. In the past, plants and animals from South America had no problem reaching Trinidad, for periodically throughout pre-history the island has been joined to the mainland, fluctuating sea levels alternately drowning and exposing the connecting 'bridge'.

LEFT About twice the weight of a large domestic cat, the ocelot is prized for its fur and is now listed as 'endangered'.

T & T have little geological affinity to the Lesser Antilles arc far away to the north. Trinidad is currently separated from South America by relatively narrow sea channels – the Dragon's Mouth in the northwest, Serpent's Mouth in the southwest and Columbus Channel in the south – while Tobago is the top of a sunken mountain chain also linked to the mainland, and today its highest point is 640 metres (2100 feet) above sea level on the 30-km (20-mile) long Main Ridge. The two islands are themselves separated by Columbus Passage, the southern opening between the Caribbean Sea to the west and the Atlantic Ocean to the east. A rocky outcrop at Galera Point, on Trinidad's side of the gap, is reputed to be the place where, in the seventeenth century, the local Amerindians threw themselves to their deaths rather than be enslaved by the Spanish.

People arrived here at least 7000 years ago, making this the earliest-settled part of the Caribbean. When the first Europeans arrived, Trinidad was called Kairi or Iere, which is often translated rather romantically as 'Land of the Hummingbird', but actually means plain-and-simply 'island'. Columbus landed here on his third voyage in 1498 and named the island the Spanish for 'Trinity', for the simple reason that he had vowed to name the next land he spotted in honour of the Holy Trinity. He called Tobago Belaforma; that is, according to one version of the story. Some historians believe that Columbus's Belaforma was really the mountains on Trinidad and that he missed Tobago altogether. It was certainly sighted later, in 1502, by Spanish explorer Alonzo de Ojeda and named Isla La Magdalena after one of his ships. This failed to stick, and the current name is thought have started out as Tavaco, the term used by the Carib Indians to describe the long-stemmed pipe in which they smoked *cohiba,* the 'weed' we know so well today. This was corrupted first to Tabagua and then to Tobago. The Spanish version, incidentally, became Tabaco, which was anglicized to Tobacco and applied to the plant that was first brought to Europe by Spanish explorers long before Sir Walter Raleigh. Today, both islands are lively and colourful and, although they have entered the industrial world courtesy of petroleum and petrochemicals, wildlife here is thriving.

People arrived here at least 7000 years ago, making this the earliest-settled part of the Caribbean

Trinidad has three mountain ranges – unimaginatively named Southern, Central and Northern, but each packed with a varied fauna and flora. The Northern Range –

geologically speaking an extension of the Andes on the South American mainland – contains the island's highest point, El Cerro del Aripo, 940 metres (3084 feet) above sea level. The mountains are clothed in ancient rainforest, and the high caves here are inhabited by the luminous lizard *Proctoporus shrevei*, named not for any ability to produce light but because its scales are highly reflective.

The El Tucuche Reserve is here too, with the island's second-highest peak, El Tucuche – 936 metres (3072 feet) – at its centre. Its forests are packed with exotic orchids and an especially large epiphyte, the giant bromeliad *Glomeropitcairnia erectiflora*, tree-top home to the golden tree frog *Phyllodytes auratus*.

ABOVE Evening at Trinidad's Nariva Swamp, the largest wetland area on the island.

This little chocolate-brown frog has, in addition to the two iridescent golden-yellow stripes that earn it its common name, a compressed head and body that enables it to live deep among the leaf axils of the plant. It differs from many other frog species in having serrated teeth and sharp 'fangs' on its jaws. The male has bigger fangs than the female, and uses them effectively in hostile encounters with other males.

In the drier forests of the Chaguaramas Pensinsula, in the extreme northwest of the island, can be found another set of formidable mandibles, those of the aggressive giant centipede *Scolopendra gigantea*, a black and shiny species that grows to over 25 cm (10 inches) long and is thought to be one of the largest centipedes in the world. The rest of the invertebrate life of the island has been little studied, but among the more noticeable forms are the arboreal chevron tarantula *Psalmopoeus cambridgei*, which has been described as 'cuddly' on account of its 'fuzzy' appearance, the fast-running and appropriately named pinktoe tarantula *Avicularia velutina*, the magnificent blue morpho butterfly *Morpho peleides* with its 15-cm (6-inch) wingspan, the recognizable trails of leaf-cutter ants *Atta cephalotes* and the formidable marching columns of army ants *Eciton* spp.

ABOVE The blue morpho butterfly has a wingspan of up to 20 cm (8 in). Microscopic scales on its wings reflect light and cause it to appear an iridescent blue.

The army ants are rarely alone, for a noisy procession of antbirds of various species follows them wherever they go.

Trinidad has abundant rainfall, the wet season occurring between June and December, but with a short, dry season or *Petit Carême* in September and October; consequently the island has substantial rivers, like the Caroni and Oropuche, that flow off the Northern Range to the sea. In the lower reaches lives the cascadura *Hoplosternum littorale*, an armoured fish with barbels under its chin that finds its way about in the murky water and gulps air at the surface, using its stomach like a lung when oxygen levels are low. Its nest consists of debris kept floating by a mass of air bubbles that the cascadura produces. It is a popular fish locally, and legend has it that anyone who eats it may wander from the island but will always return to end their days here.

The Asa Wright Nature Centre is also in the north of the island, at Arima. This is a great place for spotting birds, including the small ruby-topaz hummingbird *Chrysolampis mosquitus*, whose colours flash in the sunlight when it turns, and the tufted coquette *Lophornis ornatus*, another tiny but striking hummingbird. Birds come to the many bird feeders dotted about the site, including purple honeycreepers *Cyanerpes caeruleus* that visit special feeders resembling flowers. It is said that in a short stay you can tick off at least another 140 bird species, including some odd ones that live in an unexpected place: at Dunstan Cave, the centre has an accessible colony of 200 oilbirds *Steatornis caripensis* – extraordinary creatures that behave like bats. They roost during the day in caves and emerge at night to feed on over 36 different types of fruit in the surrounding forest. When flying inside their cave, they navigate using a primitive echolocation system and can be heard to emit clicks and raucous screams. Outside they use their eyes. They get their common English name from the fact that they were once killed and rendered down for their oil.

BELOW Leaf cutter ants are agriculturalists. They cut and collect leaf fragments that they take back to their nest in order to grow fungus gardens.

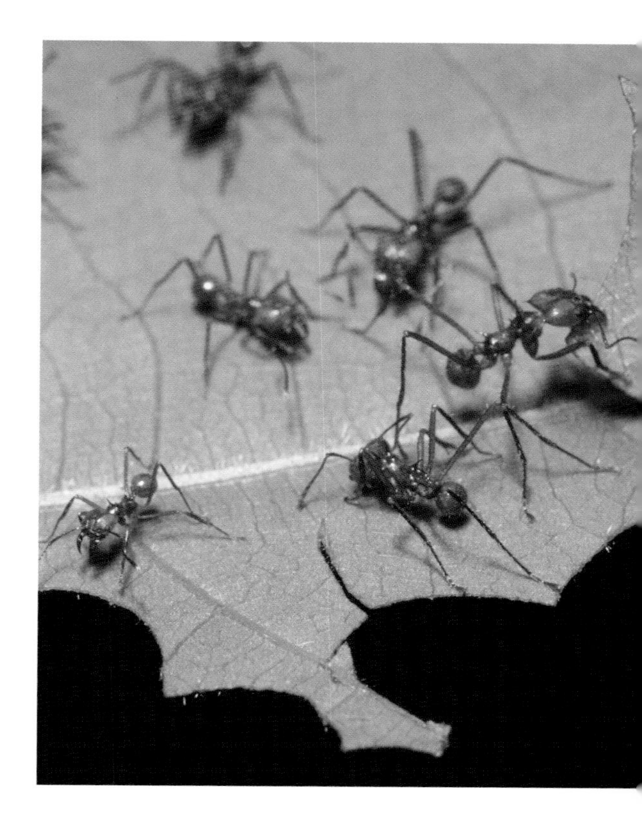

Trinidad's most unusual natural attraction is Pitch Lake, a continually replenishing lake of tar, the world's largest supply of natural bitumen and one of only three in the world. Sir Walter Raleigh used the tar to caulk his ships. The site resembles a large, constantly moving parking lot and covers 40 ha (100 acres). It can be a graveyard for some birds, but ospreys *Pandion haliaetus* are said to take advantage of the hot tar and fry their fish there.

Ospreys are just one of over 468 resident and migratory birds that have been recorded in Trinidad and Tobago, many displaying all the colours of the rainbow. In the mountains, the speckled tanager *Tangara guttata* flies around in small, loose flocks but can be seen in extended groups that may include 18 different species. In the forest, the most common parrot is the orange-winged *Amazona amazonica*. It is usually seen in pairs, but come evening hundreds may gather together in huge, swirling flocks.

At the water's edge the red-capped cardinal *Paroaria gularis* sports the colours of Trinidad and Tobago's flag, red, black and white. It perches on semi-submerged sticks and hawks for insects. Another insect-catcher is the rufous-tailed jacamar *Galbula ruficauda*, which resembles the bee-eater and similarly digs a nest burrow in sandy banks. The blue-crowned motmot *Momotus momota*, on the other hand, digs burrows up to 3 metres (10 feet) long in banks in the forest. Three species of trogon also live here, the violaceous trogon *Trogon violaceus* nesting in termite colonies.

OPPOSITE The red howler monkey is heard long before it is seen high in the canopy.

Trinidad's national bird, though, is none of these; it is the scarlet ibis *Eudocimus ruber,* many of which inhabit the Caroni Swamp, one of the largest wetland areas on the island. This bright, almost luminous red bird spends the day probing the mud, but is seen at its most spectacular at sundown when groups return to their roosts in the mangroves, decorating them like red flowers on a dark-green backdrop.

In the east, the Nariva Swamp and Bush Bush Wildlife Reserve form Trinidad's largest wetland area. A reminder of South America's proximity comes from the raucous calls of howler monkeys *Alouatta seniculus* reverberating through the trees. You hear them long before you see them. Fruit- and insect-eating white-fronted capuchin monkeys *Cebus albifrons* are here too. They are a bright bunch, capuchins having the biggest brain relative to their size of any New World monkeys.

Birds in the swamp include the savannah hawk *Buteogallus meridionalis* and the red-breasted blackbird *Sturnella militaris*. Reptiles and amphibians are represented by the 2-metre (6½-foot) long spectacled caiman *Caiman crocodilus*, the green anaconda *Eunectes murinus,* a species known to grow up to 10 metres (30 feet), and the Surinam toad *Pipa pipa*, which carries its tadpoles in pits in its back. All share open-water sites with a small population of West Indian manatees *Trichechus manatus*.

The island's Central Range has caves, the most important for wildlife being those near the top of Mount Tamana. The caves, at 230 metres (750 feet) above sea level, harbour 11 species of bats; an estimated 1.5 million individuals roost and breed here. Piled on the cave floor is bat guano, the natural if unsavoury home of millions upon millions of cockroaches.

At the coast, Trinidad's beaches host nesting sea turtles. Matura Bay is particularly important, for each night during May through July female leatherback turtles *Dermochelys coriacea* haul their enormous bulk to the top of the beach, where they deposit their eggs in a pit dug in the sand. As many as 150 females can pitch up in a single night, some having swum from feeding grounds as far afield as the west coast of Scotland. Each is up to 1.8 metres (6 feet) long, which makes the leatherback the largest species of turtle in the world.

The appropriately named Turtle Beach and nearby Grafton Beach Resort are also turtle beaches but, like many of the Caribbean islands, Tobago is becoming better known worldwide for its dive sites. One of the most popular is the wreck of the

ABOVE The green anaconda, with a maximum weight of 250 kg (550 lb) and a girth up to 30 cm (12 in) across, is considered to be the heaviest snake in the world but not the longest.

OPPOSITE The spectacled caiman gains its common name from the bony ridge between the eyes that gives the appearance of spectacles.

In the east, the Nariva Swamp and Bush Bush Wildlife Reserve form Trinidad's largest wetland area

ABOVE The white-fronted capuchin prefers trees over 30 metres (100 feet) high where it feeds on fruits, leaves, berries, nuts, gums and bark, but will also take insects, birds' eggs and baby mammals.

Maverick ferr, that in 1997 was deliberately sunk in 30 metres (100 feet) of water to create an artificial reef off Rocky Point. The wreck has had all its doors and windows taken out so that divers can explore inside. Two of the fish that have taken up residence are a 1.2-metre (4-foot) long jewfish *Epinephelus itajara* and a giant barracuda *Sphyraena barracuda*.

Currents around Tobago are strong, so the island is popular for 'drift diving', especially in a particularly fast current off Speyside known as the African Express, which flows at a regular 4 knots (7.5 km/h or 5 mph). One of the natural features over which divers drift is the biggest brain coral *Diploria labyrinthiformis* in the western hemisphere. It is to be found at Speyside's Kelliston Drain, and is 3 metres (10 feet) high and over 5 metres (16 feet) wide.

Drift diving in this area relies partly on the outflow of the Orinoco River, which floods from July to September, changing the colour of the water from azure to greenish-yellow. Sometimes, two distinct layers form at the surface: a thick, green, warm, mainly freshwater layer above a clear saltwater base, giving divers the impression of being under a vast canopy. A bloom of algae is responsible for the green colour, and at Speyside a 'flight' of giant, filter-feeding manta rays *Manta birostris* arrives each year to take full advantage of the glut of microscopic food.

A short boat trip from Speyside is Little Tobago, a bird sanctuary with red-billed tropicbirds *Phaethon aethereus* nesting on the scrubby cliffs and all manner of

seabirds – brown boobies *Sula leucogaster*, bridled terns *Sterna anaethetus*, sooty terns *Sterna fuscata* – nesting near by. Magnificent frigatebirds *Fregata magnificens* and red-footed boobies *Sula sula* nest on the adjacent island of St Giles.

Tobago and its smaller islands are some distance from the Orinoco and less prone to changes in its flow, but Trinidad is directly opposite the river's vast delta, and when its coastal waters are swamped by the Orinoco flood, the decreased visibility of its sediment-filled water helps to hide the newborn of a particularly attractive shark – the golden or small-eye hammerhead *Sphyrna tudes*. It is no more than 1.5 metres (5 feet) long when fully grown and, unlike the normally grey above and white below hammerhead species, it sports a yellowish-golden hue. The colour is due to its diet, for juvenile sharks eat penaeid shrimps *Xiphopenaeus kroyeri,* and adults consume marine catfishes of the family Ariidae and their eggs, all of which contain a carotene-like pigment.

The Orinoco itself opens to the sea on the Caribbean coast of Venezuela in a great labyrinth of water channels – the Orinoco Delta. Columbus came here on his third voyage in 1498 and today, just as then, when the river is not in flood, it is a complex of moist forest and wetlands the size of Wales or New Jersey, one of the largest intact areas of wetland remaining on Earth. Here, red brocket deer *Mazama americana* are alert for jaguars *Panthera onca*, and there are bush dogs *Speothos venaticus* and crab-eating foxes or zorros *Cerdocyon thous*, coatis *Nasua nasua* and kinkajous *Potos flavus*, jaguarundis *Herpailurus yagouaroundi* and ocelot. Saki monkeys *Pithecia* spp. share the trees with weeping capuchins *Cebus olivaceus* and howler monkeys. This is a South American wildlife paradise, the only interlopers being the Warao people who build their houses beside the channels, hunt in the forest and fish in the river.

Birds include the Orinoco goose *Neochen jubata* that rarely swims or flies but nests in hollow tree trunks, such as those of the 35-metre (115-foot) tall moriche palm *Mauritia flexuosa*, whose fruit is used to make jams and fermented wines. The pulp also contains high levels of vitamin A and is used to treat burns.

Ducks and geese are well represented, with several species of whistling or tree ducks. The black-bellied tree duck *Dendrocygna autumnalis* has a red bill and legs, while the white-faced tree duck *D. viduata* has the obligatory white face but rarely perches in trees, so its alternative name 'white-faced whistling duck' is more appropriate; it is especially noisy at dawn and dusk, emitting different whistles for different behaviour.

Anhingas or snakebirds *Anhinga anhinga* are often seen catching fish in the river channels. Their plumage is not waterproof like that of other water birds, an adaptation that enables them to achieve almost neutral buoyancy, helping them to stay underwater for long periods while pursuing fish. When resting between hunting excursions, they stand on logs or branches with their wings spread open to dry.

The river, though, rises and falls with the wet and dry seasons, and when the floods come it is all change. Rodents such as agoutis *Dasyprocta aguti* and pacas *Agouti paca* that forage on the forest floor during the dry season are replaced by the Orinoco crocodile *Crocodylus intermedius* and fish such as the meat-eating piranha *Pygocentrus nattereri* and vegetarian pacu *Colossoma macropomum*. Some animals can cope with the changes. Capybaras *Hydrochoerus hydrochaeris*, the world's largest rodent, and giant river otters *Pteronura brasiliensis* live here all year round.

To the west of the Orinoco, where Venezuela faces the Caribbean Sea, there are 400 km (250 miles) of narrow coastal strip dotted with isolated stands of mangroves, considered to be the most pristine mangrove forests in the Americas. Many of the stands are small and widely separated, but between the Gulf of Paria and the Orinoco Delta is the largest, with mangroves reaching heights of 35–40 metres (115–130 feet). Here, the jungle meets the sea in the Parque Nacional de Peninsula de Paria, the only home on Earth to the endangered yellow-faced or paria redstart *Myioborus pariae* and scissor-tailed hummingbird *Hylonympha macrocerca*.

Care should be taken when looking up, however, for down on the forest floor are extremely venomous snakes. The bushmaster *Lachesis muta* grows to 3 metres (10 feet) or more, making it the longest venomous snake in the western hemisphere. It is also considered one of the most dangerous in the world. Although its venom is similar in strength to that of a diamond-back rattlesnake *Crotelus atrox*, it pumps enormous quantities into its victims. Even during treatment, 80 per cent of human casualties die. A second hazard is the fer-de-lance *Bothrops atrox*, a fast-moving night hunter that frequents stream banks, ditches and plantations. As a member of the pit viper family like its relative in St Lucia (see chapter 4), it recognizes the heat profile of its warm-blooded prey using heat-sensitive pits behind and above the nostrils. It strikes in less than the blink of an eye. It is also on the 'top ten' dangerous list, accounting for more human deaths than any other reptile in the Americas.

Less dangerous and far less scary are the mangrove tree crabs *Aratus pisonii*, leaf-consuming mangrove crabs *Ucides cordatus* and tree-climbing red mangrove crabs *Goniopsis cruentata* that clamber about in the tangle of branches and roots. The red mangrove crab keeps itself clean with a clever trick. It first produces foam from its gill chambers and then, using the froth like soap, it rubs together its dirty legs and mouth appendages. When it dries, the crab's exoskeleton is clean and glossy, and any growths, such as algae, have disappeared. Its main predators are

OPPOSITE Tall, arching prop roots identify the red mangrove. They are also located closer to the water than other mangroves due to their high salt tolerance.

the crab-eating raccoon *Procyon cancrivorus*, the rufous crab-hawk *Buteogallus aequinoctialis* and people who harvest the crabs, especially *Ucides*, for food; in fact, *Ucides* is probably the most exploited animal in these mangroves.

Further again to the west is the nationally famous wildlife refuge Parque Nacional Henri Pittier. Established in 1937 and named after a highly regarded Swiss botanist, it is Venezuela's oldest reserve. It extends from the sea up to Cenizo Peak at 2430 metres (7972 feet), where the Paso Portachuelo is a natural route for migrating birds and insects. Vegetation includes the giant and endemic cucharón or niño tree *Gyranthera caribensis*, which grows to 60 metres (200 feet) tall, and the giant-leaved *Gunnera pittierana*. The fauna includes the unusual marsupial frog *Gastrotheca ovifera*, which gets its common name from the way in which the female of the species carries her developing eggs in a pouch on her back, a unique form of parental care among frogs.

And, while exotic mammal species such as jaguars, pumas, tapirs, armadillos and a great variety of New World monkeys are present, the park is very popular with birdwatchers, for 40 per cent of Venezuela's and 7 per cent of the world's bird species are found here – a staggering 580 of them. Parrots, parakeets, parrotlets, hawks, owls, curassows, guans, toucans, oropendulas, tinamous, antthrushes, bellbirds, woodpeckers, manakins, caciques, chachalacas may all be seen – an extraordinary assemblage of South America's birds.

The further west you travel, the hotter and drier the climate becomes, and at Médanos de Coro, an extensive area of sand dunes resembles a North African desert. One site of interest is Mount Chichiriviche, into whose slopes is gouged the so-called Cave of the Indian, a 75-metre (250-foot) deep sinkhole that contains petroglyphs dated 3400 BC, the writings and drawings of the Caquetios people. In another cave near by are the statues of Virgin del Valle, the patron saint of Venezuelan fishermen. Each July a procession of fishing boats passes by and is blessed by the local bishop.

Venezuela's coastal strip extends westwards to the Laguna de Cocinetas and includes a colony of 15,000 Caribbean flamingos *Phoenicopterus ruber ruber* in Los Olivitos Swamp, one of only four sites in the Caribbean with a substantial nesting population. The area could be labelled 'flamingo junction', for not far away is the mangrove-lined Laguna de Tiraya. Located on the Paraguaná Peninsula, the most northerly part of mainland Venezuela, it is a feeding site for flamingos that nest not on the mainland itself, but on the offshore islands.

The most easterly and remote of these is Los Testigos, meaning 'the Witnesses', location of deserted sandy beaches and sand dunes more than 100 metres (330 feet) high. The entire place is run by fishing families who guard their underwater wilderness closely – no diving or spear fishing permitted, only snorkelling – and who must rarely stroll inland, for this is the realm of tiny black flies, clouds of them.

The largest of Venezuela's offshore islands is Isla de Margarita, famous for pearls that, in the sixteenth century, represented a third of the total New World tribute to the Spanish crown. The north and west of the island are mainly wetland and sand dunes, home to the rare yellow-shouldered parrot *Amazona barbadensis*. The human population is crammed into the southeast.

Nowadays, one of Margarita's main claims to fame is its wind, especially at Playa El Yaque between January and June. Such is the shape of the bay here – solar-heated shallow water and nearby mountain range – that the wind accelerates as if going through a funnel, the so-called Venturi effect. The result is constant side-shore winds of 25 knots (65 km/h or 40 mph) or more, bliss for windsurfers.

To the west of Margarita is the Parque Nacional Archipiélago de los Roques, which protects 55 coral islands and countless sandy cays and islets. The main vegetation is cactus and mangrove; turtles and iguanas nest on the deserted beaches; and 92 species of birds spend at least part of their year here, many heading for North America to breed in the summer. Fishing for bonefish or *pez ratón Albula vulpes* is permitted, and some anglers reckon the archipelago is one of the best sites in the world for this species. The fish are caught, released and caught another day.

The coral reefs are still intact, with different species of reef-building hard corals at various depths. Outside the ring of islands, elkhorn and staghorn corals dominate a narrow ledge of shallow-water reef. The sloping drop-off has patches of star and brain corals at first, but the deeper you go the more the plate corals, black corals and elephant-ear sponges take over. Rocky spires covered in cup corals and sponges are meeting places for queen angels *Holocanthus ciliaris* and red-lipped blennies *Ophioblennius atlanticus*. Submarine caves hide clouds of tiny fairy basslets *Gramma* spp., while giant manta rays and spotted eagle rays *Aetobatus narinari* are attracted to the reef's outer edge.

Isla de Coche and Isla de Cubagua, between Margarita and the mainland, featured in Columbus's log. It was here that he was first enthralled by the region's pearls; even today pearls the size of pigeons' eggs are still to be found, albeit illegally. Near the reef at Los Palanquinos on the Isla la Tortuga, about 80 km (50 miles) west of Margarita, the water is so clear that a diver can see 20 metres (60 feet), and a boat anchored in 2 metres (6½ feet) of water looks as if it is floating on air. The island is deserted except for a few fishing camps.

Even more remote are Isla la Blanquilla, Isla la Orchila and Isla de Aves, refuges for pirates in a bygone age and now pristine sanctuaries for wildlife. Isla la Blanquilla is no higher than 21 metres (70 feet) above sea level and is surrounded by alabaster-white sand beaches (from which it gets its name), low cliffs, sea caves, natural arches and reefs with black corals. It sits close to a deep-water trench, where just 20 metres (65 feet) from the shore the sheer submarine wall plummets straight down over 900 metres (3000 feet). The island itself is covered with cactuses and is home to the introduced yellow-shouldered parrot, feral donkeys and a species of burrowing owl.

The untouched archipelago of Isla de Aves has underwater life that on other

Caribbean islands is extinct or almost so, like the queen conch or botuto *Strombus gigas*. It is the biggest mollusc with a shell to be found in the Caribbean: up to 40 cm (15 inches) long and weighing 3 kg (6½ lb). Its eggs are also remarkable. They are laid in strings that can be up to 23 metres (75 feet) long. The problem for the conch, however, is that it tastes good, and consequently it has been collected almost to the point of oblivion. It is practically extinct in the Grenadines, Puerto Rico and Florida Keys, but is still abundant in Las Aves and on Los Roques and Isla la Orchila. Piles of empty conch shells, however, indicate that even here its future is uncertain.

One island in the archipelago, Las Aves de Barlovento (meaning 'windward'), has enormous bird colonies – gulls, pelicans and boobies – and the trees in the mangrove channels grow to 6 metres (20 feet) tall. Las Aves de Sotavento (meaning 'leeward') is wreck heaven, the latest discovery being one of seven French ships from the fleet of Count d'Estrées that all went aground on 4 May 1678. Wreck divers are searching for the other ships and the treasures they contain. Isla del Tesoro, meaning 'Treasure Island', received its name in the early nineteenth century after fishermen found traces of treasure there.

To the west of the Venezuelan island dependencies is the southern half of the Netherlands Antilles – Bonaire, Curaçao and Aruba, known locally as the ABC islands. They are exposed, windblown, arid and desert-like, with carelessly strewn red granite boulders (the debris from an ancient tsunami) and cactuses the size of trees. There are no mountains and little rain, and the wind is so strong all the trees bend one way.

Bonaire is best known for its fringing reef. With no rivers to bring down sediment that could smother the corals, it is in excellent shape and is being kept that way, making it one of the best marine sites in the entire Caribbean. Catching sea turtles was made illegal in 1961, when the rest of the world still sipped turtle soup. In 1971, spearfishing was banned, and in 1975 it became an offence to break, remove or sell coral. In 1979, the Bonaire Marine Park was established, protecting wildlife from sea level down to 60 metres (200 feet), and so today the reefs are in pristine condition.

One of the best snorkelling sites is Playa Bengé, where a diver swimming directing away from the beach will pass over alternating rows of coral and sand, the haunt of jewelfish *Hemichromis bimaculatus*, mahogany snappers *Lutjanus mahogoni* and huge tiger groupers *Mycteroperca tigris*. Dive sites throughout the island are marked with yellow stones, the name of the site written on each stone, and it is said that if a fish exists in Caribbean waters, you will find it at Bonaire.

The untouched archipelago of Isla de Aves has underwater life that on other Caribbean islands is extinct

In the northeast of the island is the Washington-Slagbaai National Park. In the northern part of the park vegetation is dominated by two species of cactus in which yellow-shouldered parrots and parakeets nest. The yatu *Lemaireocereus griseus* has spines in rosettes while the dense spines of the kadushi *Cereus repandus* stick out in all directions; the former grows straight and the latter is branched. Smaller cactuses nearer the ground have blue whiptail lizards *Cnemidophorus murinus ruthveni* living on, in or under them. The lizards fight over and feed on the fruit of the Turk's head cactus. Parrots, orioles and mockingbirds depend on cactus fruits too, and, in this dry environment, water is at a premium, so iguanas will also battle for the possession of a spring.

ABOVE Refurbished huts in the northern part of Bonaire were once the homes of slaves working the salt pans.

In the eighteenth century, the area was important for salt extraction. Dotted about the landscape are the huts where slaves who did the extracting once lived. Lurking near by are often the feral donkeys that were used to carry the salt. Today, from the top of Mount Brandaris, the island's highest point at 241 metres (790 feet), the mountains of white salt belonging to the company Cargill Incorporated can still be seen to the south. They are known affectionately as the 'Bonairian Alps'.

Many of the bone-white saltpans or *salinas* today are the breeding and feeding grounds of Caribbean flamingos, which filter water snails, brine flies and shrimps from the salt water. Goto Meer in the park is a known flamingo site, as is the Pekelmeer Flamingo Sanctuary in the south of the island; together they account for 7000 breeding birds, making Bonaire the largest natural flamingo sanctuary in the western hemisphere.

The southern part of Washington-Slagbaai is fenced to keep out feral goats, so saplings have a chance to grow. The soil here is more fertile than in the north, so the vegetation is lush and includes the mesquite tree or indju *Prosopis juliflora*. At one time, its branches were cut to produce charcoal, but it was never overexploited. The tree was cut to within about 30 cm (12 inches) of the ground, and in five years it grew back to its former size. In this way, Bonaire's trees were never lost.

Aloe *Aloe barbadensis* was cultivated for its medicinal properties, and the gnarled and twisted Brazil wood tree *Haematoxylon brasiletto* for the red dye that can be extracted from the wood. In fact, sixteenth-century maps label Bonaire as *Yslas de Brasil*, meaning 'Island of the Brazil Tree' – one reason the Dutch seized it from the Spanish in 1636. The mata-piskà tree *Jacquinia barbasco* was also of value. The local Amerindians threw its branches, berries and leaves into the water, and the chemicals that leached out paralysed the gills of fish. The fishermen then gathered up the anaesthetized fish with ease.

Throughout Bonaire birds are never far away; about 90 species are migrants, heading to and from North or South America, making this and the other islands of the Netherlands Antilles stepping stones between continents.

About 15 minutes by air to the west is Curaçao, a bigger island than Bonaire, but divided between an eastern suburban side, dominated by cruise ships and the oil industry, and a wild west. As on Bonaire, the diving on the west coast is superb. One site – called Seldom, because it is often rough and seldom visited – has a steep wall close to the shore where shoals of barracuda slowly pass by and giant green moray eels *Gymnothorax* spp. swim in the open. Then there is Airplane Wreck, with a ditched plane marked by a buoy, and Barracuda Point, where inquisitive but intimidating barracuda will actually follow divers about. The national Underwater Park is over 20 km (12 miles) long, and Klein Curaçao, a small island off the east coast, is a prime dive site. About 50 species of reef-building corals contribute to the steep-profile fringing reefs, which drop to a depth of 40 metres (130 feet) just 150 metres (500 feet) from the shore.

There are 38 beaches around the island, some backed with towering sea cliffs, which provide one of the island's entertainments – high divers who launch into space from cliffs at Westpunt Beach, named for its position at the westernmost point of the island. More restrained visitors are the sea turtles that come ashore to deposit their eggs, especially at protected beaches such as those at Shete Boka Park. At Boca Pistol enormous waves crash into the rocks with the sound of a gun firing. Elsewhere, there are large sea caves carved out by the pounding surf.

To the north of Willemstad, Amerindians once used the spectacular Hato Caves for shelter and as a burial place. Their petroglyphs, estimated to be 1500 years old, decorate the walls and ceilings. The skeletal remains of a family were once found here, their skulls grouped together with an adult male at the centre. In the seventeenth and eighteenth centuries escaped slaves used the caverns to hide from the plantation owners. Today, a colony of long-nosed bats *Leptonycteris curasoae* has made its home here, and the passages are now walked by paying visitors who pass an extraordinary wealth of stalagmites, stalactites and other cave formations with such descriptive names as Pirate's Head, Sea Tortoise and the Madonna.

Outside, the wildlife on Curaçao is protected in the Christoffel National Park, a nature reserve created from three former plantations. At its heart is Mount Christoffel – 375 metres (1230 feet) – where yellow warblers *Dendroica petechia*, crested caracaras *Caracara plancus* and white-tailed hawks *Buteo albicaudatus* can be seen, and crested bobwhites *Colinus cristatus* breed close to the car park.

Throughout Curaçao, 168 bird species have been recorded, of which at least 51 breed on the island, 71 are migrants from North America, 19 come from South America, and 19 are seabirds. Two subspecies are unique – a parakeet *Aratinga*

pertinax pertinax and a barn owl *Tyto alba bargei*, both endangered. Mammals are represented by 11 native species: the Curaçao white-tailed deer *Odocoileus virginianus curassavicus*, a mouse *Baiomys hummelincki*, a cottontail rabbit *Silvilagus floridensis nigronuchalis* and eight bats. Research has shown the importance of the island's nectar-feeding bats – the lesser long-nosed *Leptonycteris curasoae* and the long-tongued *Glossophaga longirostris* – as the principal pollinators of the large, columnar cactuses. The cactus flowers produce most of their nectar to coincide with times when bats are on the wing. If the bats should ever disappear, the cactuses would go too, for they are not self-pollinators.

ABOVE There are many subspecies of the white-tailed deer throughout the Americas, and the island of Curacao has its very own.

Notable plants in Christoffel include four species of orchid. The most common is the purple-flowered banana shimaron *Schomburgkia humboldtii*, which is an epiphyte on trees and can grow to a clump over a metre (3¼ feet) across, some stems reaching 4 metres (13 feet) long and sporting 15–20 flowers. After rain, a large group can get so heavy that it crashes to the ground, bringing part of the tree with it. The *dama di anochi* orchid *Brassavola nodosa* is also found in trees, often in sites where they are dusted by salt spray from the sea. A small yellow-flowered orchid *Polystachya cerea* is found only in the more remote and isolated areas of the park, and *Epidendrum atropurpureum* is probably now extinct. This white-flowered, red-hearted beauty was discovered by a local cleric, and its location, a single rock, was kept a very tight secret. It flowered as regularly as clockwork every February. It was checked frequently, and in 1970 a second plant was found in a nearby tree. After the rains in 1971, however, the tree fell over, and goats ate the orchid. Ten years later, orchid thieves discovered the location and stole the orchid on the rock.

Orchids, coral reefs and cottontails aside, excitement among the island's naturalists reached fever pitch in October 2005, when a West Indian manatee turned up on the north coast, 35 years after one was last spotted here. It is thought to have arrived from the Venezuelan coast, about 65 km (40 miles) away to the south. The locals hope it will stay and that it is one of several manatees that have been hanging out on this deserted coast, remaining undetected all this time.

ABOVE Blue whiptail lizards are common at Playa Funchi on Bonaire, where they will take breadcrumbs from out of your hand.

About 65 km (40 miles) to the west is the third of the ABC islands. Aruba, meaning 'red gold', is outside the hurricane belt, but its blistering heat is soothed by trade winds all year round. It is a basically flat island with a few low, rolling hills, its highest point being Mount Jamanota, 188 metres (617 feet) above sea level. The mountain is part of the Arikok National Park, a protected area that accounts for almost 20 per cent of the island. In the park can be found strange forests of cactuses, some the size of pencils and others as big as telegraph poles, and dotted here and there on exposed ridges are wind-sculpted divi-divi or watapana trees *Caesalpinia coriaria*, which always stretch to the west. The divi-divi's seedpods contain large quantities of tannin, so they were at one time exported to tanneries in the Netherlands.

The park holds many of Aruba's indigenous animals, including two endemic snakes, a lizard and two species of birds. One snake is the venomous cascabel *Crotalus thurissus uni-color*, also known as the island's rattlesnake but one that does not use its rattle. The other is the non-venomous santanero or Aruba cat-eyed snake *Leptodira bakeri*, which if you pick it up has the unattractive habit of defecating in your hand. The lizard is the endemic kododo blauw or Aruban whiptail *Cnemidophorus arubensus*, and the birds are the shoco or Aruban burrowing owl *Athene cunicularia arubensis*, which can be seen standing at its nest entrance from the sixth hole on the Tierra del Sol golf course in the north of the island, and the green-bodied and yellow-headed prikichí or Aruban parakeet *Aratinga pertinax arubensis*.

Other birds include the tropical mockingbird *Mimus gilvus*, the bright orange but shy trupial *Icterus icterus* and the barika geel or banaquit *Coereba flaveola*, a small yellow-bellied bird that satisfies its sweet tooth with spilled sugar from the breakfast table. As there are no rivers on the island, many birds are attracted to the Bubali Ponds, former saltpans and now freshwater lakes topped up by a water-treatment plant, where scarlet ibis and many water birds such as pelicans and cormorants can be seen. A major nesting site is San Nicolas Bay Keys in the southeast, where sooty terns nest cheek by jowl with black noddies *Anous minutus*, and brown noddies *Anous stolidus* nest in or under buttonwood trees *Platanus* spp.

Aruba is punctured by numerous caves, once the undoubted hiding places of pirates and buccaneers. Guadirikiri Cave has two chambers where the light shines through from the surface, while Fontein has petroglyphs on the ceiling, and Huliba is known as the 'tunnel of love' on account of its heart-shaped entrance.

In the northeast is a desert-like landscape of enormous sand dunes, but the most interesting geological features are the Ayo and Casibari rock formations. Just to the north of Hooiberg, vast monolithic quartz-diorite boulders are scattered across the landscape, and at Ayo huge stones stacked one of top of the other were once the dwellings of an ancient people who left petroglyphs for us to puzzle over today; how the boulders got to where they are is a puzzle in itself.

Aruba also once had a famous natural arch, the pride and joy of the local people and much photographed by visitors; but alas, if you want to see it you are too late: on 2 September 2005, it collapsed into the sea.

6

WESTERN CARIBBEAN

WHILE ALL the countries visited so far are influenced by the ocean circulation and weather systems of the Atlantic, those in the western part of the region have two coasts – one facing the Caribbean Sea and the other the Pacific Ocean. The two sides are separated by a spine of highlands, with some of the most active volcanoes in the world bordering the Pacific, the so-called 'ring-of-fire'. The Caribbean side, by contrast, is mostly flat coastal plain with forests, mangroves, islands and coral reefs.

PREVIOUS PAGES The chestnut mandibled toucan is found in moist forests from Honduras to Colombia, where they are seen flying in groups of three to twelve birds.

Colombia has been long famed for its coffee and nowadays for drugs, guerrillas and kidnappings; it is a country in turmoil, but it also has spectacular wilderness areas and is extraordinarily rich in wildlife. About 1770 species of birds have been recorded here, the most of any country in the world, and 130,000 species of plants collected. Behind the political confusion and everyday violence hides an untapped tropical paradise.

Part of that paradise can be found along Colombia's Caribbean coast where, in addition to its wildlife, features at least one astonishing archaeological site.

The coastal lowlands are contained in an imaginary triangle whose base runs from the Golfo de Urabá to the Venezuelan frontier and whose longest side forms the coastline. Immediately to the west of the border is the Guajira Peninsula, at whose northeastern tip is an arid region containing the Parque Nacional Natural de Macuira. Cerro Palúa, at 865 metres (2837 feet), is the highest point in an isolated range of hills, an oasis in a semi-desert and home to an unexpected biological curiosity. Here, in a stand of tropical forest, exists a species of poison-dart frog

ABOVE Sunset over distant islands in the San Blas Archipelago, Panama.

Colostethus wayuu that lives nowhere else. This little character, no bigger than your thumbnail, secretes toxins from its skin that at best can leave a bad taste in a predator's mouth and at worst will kill it. The male of the species has another unexpected attribute: his spermatozoa have two flagella rather than the usual single one, a trait shared with only two other types of frog.

Further to the west, about 25 km (15 miles) from the city of Riohacha, is the Santuario de Fauna y Flora Los Flamencos, two large, saline lagoons and adjacent marshes that are separated from the sea by sand bars. This, as its name suggests,

is a sanctuary for hundreds of Caribbean flamingos *Phoenicopterus ruber ruber* that nest here, each on its own 60-cm (2-foot) high mound of mud. The birds are part of the huge, mobile population that migrates to and from the Netherlands Antilles and Venezuela. Other species here all year round include pelicans, herons, gulls, cormorants and the nacunda nightjar *Podager nacunda*, a bird so well camouflaged that it is very hard to see against its stony nest-site background.

To the southwest of Riohacha, the Sierra Nevada de Santa Marta is in another isolated range, high but quite separate from the northern end of the Andes. The mountains rise abruptly from the edge of the Caribbean Sea to snow-capped peaks just 45 km (30 miles) away. This is the highest coastal mountain range in the world and the location not only for Colombia's highest summits – Pico Cristóbal Colón and Pico Simón Bolívar, both 5775 metres (18,950 feet) above sea level – but also for the extraordinary ruins of La Ciudad Perdida, the Lost City. La Ciudad is pre-Columbian, possibly built as early as the fifth century AD, and rivals Machu Picchu for its spectacular position high on the forested slopes of the steep-sided Cerro Corea.

The place was found in 1972 by treasure looters, and the outside world was not alerted until 1975, after gold figurines and ceramic pots began to appear on the black market. Local tribes knew about it, but they had kept it to themselves, believing it to be at the heart of a group of towns and villages inhabited by their

ABOVE A stocky, muscular body and shorter tail distinguishes the New World jaguar from its Old World relative the leopard.

forebears, the Tairona. Up to 8000 people are thought to have been housed here, but the site was abandoned shortly after the Spanish arrived, when the Tairona escaped into the forest. It consists of about 170 platforms and terraces, large, round plazas and a network of paved footpaths, and is reached from the forest below by a flight of over 1200 steps and hand-carved tunnels. The Tairona designed it so that they could hear visitors approaching and take appropriate action, depending on whether they were friend or foe. Today's visitors have other things to worry about, for the site is in the middle of fighting among right- and left-wing guerrilla groups and the Colombian army. Currently, a paramilitary right-wing group is the self-claimed protector of the city, and, since 2005, visitors have been able to enter the area once more.

If a five-day trek, over 1200 steps and the prospect of being kidnapped are not enticing, then a scaled-down version of La Ciudad Perdida can be seen in a valley between mountain peaks at El Pueblito in the Parque Nacional Tayrona. The park itself borders some of the north coast's best beaches, and scattered all around are granite boulders, some in piles that hide small caves and rocky passageways, others covered with vines and creepers. Inland, reintroduced Andean condors *Vultur gryphus* are occasionally spotted in the distance, soaring on thermals between mountain ridges. In the forest, the delightful shining-green hummingbird *Lepidopyga goudoti* can put in a brief appearance, but a glimpse of a puma *Puma concolor* or jaguar *Panthera onca* among the trees is unlikely ... although they are there, somewhere.

Adjacent to the Santa Marta mountains is the Ciénaga de Santa Marta and the estuary of the Magdalena River, the most important wetland and largest mangrove area on Colombia's Caribbean coast, and a UNESCO Biosphere Reserve. Here, in a system of lagoons and delta channels, salt water and fresh intermingle, corals and mangroves thrive side by side, and mud banks in the Magdalena delta come and go. The dry land areas are covered by very dry tropical forest. Trees such as the moro or fustic tree *Chlorophora tinctoria* grow here. Dyes are extracted from its bark and its latex is used as a painkiller. The Madras thorn *Pithecellobium dulce* is here, too, with its distinctive seedpods shaped like the human ear. They have red pod valves, black seeds and a white fleshy aril, which not only attracts birds as seed dispersers but is also eaten as sweets or *dulce* by children.

The mangroves have the mangrove oyster *Crassostrea rhizophorae*, a delicacy appreciated by tourists and local oystercatchers *Haematopus palliatus* alike. Its pearls also have value. When Sir Francis Drake was in the area in 1586, he 'acquired' a very large pearl that he presented to Queen Elizabeth I of England and that was set in her state crown. This is an important area also for commercially exploited fish such as mullet and shad, and some of the fishermen

A glimpse of a puma or jaguar among the trees is unlikely ... although they are there, somewhere

live in villages built on stilts. Competing with fishermen for food and living space are the narrow-snouted spectacled caiman *Caiman crocodylus* and American crocodile *Crocodylus acutus*, the latter's meat and skin making it a target for poachers and putting it at risk of extinction in Colombia.

At the northern end, the lagoon complex is separated from the Caribbean by the barrier island Isla de Salamanca, which has an inlet on its eastern end that connects the largest lagoon, Ciénaga Grande, directly to the sea. It is a popular stop-off point for migrants – herons, ducks and ibis, including the whispering ibis *Phimosus infuscatus* and, increasingly, since the late twentieth century, the glossy ibis *Plegadis falcinellus*. Birds exclusive to the area include the bare-throated tiger-heron *Tigrisoma mexicanum* and the sapphire-bellied hummingbird *Lepidopyga lilliae*, another local at serious risk of extinction.

Birds, however, are not the only attractions along this stretch of coast. There is some attention-grabbing geology too. Between Santa Marta and the Gulf of Urabá in the west, there are several locations where visitors can experience so-called 'mud volcanoes'. At Arboletes, for example, a circular mud lake about 30

ABOVE The glossy ibis probably crossed the Atlantic from Africa to Latin America in the 19th century.

metres (100 feet) across contains warm grey-black mud permeated by bubbling sulphurous gases in which visitors can wallow. At some sites, like the mud volcano of Volcán del Totumo near Cartagena, a crater-like ridge of hardened mud has built up around the lake, pushing the mud bath up to 25 metres (80 feet) above the surrounding countryside. Their origin is not clear, although there seems to be some link between the location of the north coast of Colombia, close to the boundary between the South American and Caribbean continental crustal plates. It is thought that heated gases from deep down percolate up through the sediments, and these bubble to the surface as mud lakes.

To the west of the Gulf of Urabá is the Darién Gap, an impenetrable forest and swampland area that Colombia shares with neighbouring Panama. To see it today, you wonder at the absurdity of attempting to establish a settlement here, but in the late seventeenth century that is precisely what happened.

It was here that Scottish venturer William Patterson, who made his fortune as a founder member of the Bank of England, tried to set up a colony in the hope of transporting goods across land from the Pacific to the Caribbean coast, thus

ABOVE A stream cuts through Panamanian tropical rainforest.

cornering the lucrative Pacific trade market. What he and his companions found when they arrived in 1698, however, was quite different from what they had expected. The land could not be farmed, the local Amerindians were not interested in the trinkets they had brought to barter, and torrential rains brought diseases. One of the many was *vinchucas* or Chagas' disease, caused by the parasite *Trypanosoma cruzi* and spread by blood-sucking triatomine bugs or 'kissing bugs'. At one point ten people a day were dying. The final straw was the news that the Spanish were about to attack the survivors, so the British contingent fled – less than a year after they had landed. Only 300 out of the 1200 who originally set out from Edinburgh returned home alive.

And that was not the end of the story, for another expedition, with 1302 settlers, set out for the Darién in 1699, unaware that their countrymen were returning. They rebuilt the colony but were attacked mercilessly by the Spanish and very few made it back to Scotland. Patterson's dream had to be abandoned; he had been beaten by this desolate place ... and, of course, by the conquistadors.

On the Panama side of the Darién Gap, so-called because there is a break here in the Transamerica Highway, is the Darién National Park, a bridge between two continents and another UNESCO Biosphere Reserve. The monsoon forest has trees growing up to 50 metres (165 feet) tall, the most common being the cuipo *Cavanillesia platanifolia*, a giant with an unusually straight trunk, no buttress roots and enormous multi-winged fruits. All manner of South American mammals live in the park, from bush dogs *Speothos venaticus* to giant anteaters *Myrmecophaga tridactyla* and Baird's tapir *Tapirus bairdii* to brown-headed spider monkeys *Ateles fusciceps*. The spider monkeys are fruit-eaters and need large tracts of healthy rainforest in which they can search for trees as they come into fruit. They have a prehensile tail, like a fifth limb, that enables them to move rapidly through the forest.

Cerro Pirre, a highland wilderness and the country's most productive bird-watching site, has populations of the golden-headed quetzal *Pharomachrus auriceps* and saffron-headed parrot *Pionopsitta pyrilia*, but top bird in the park is Panama's national bird, the harpy eagle *Harpia harpyja*, which can snatch a monkey from a branch in the blink of an eye.

BELOW A bush dog emerges from its den, which it can share with up to ten other adults.

Alongside the Darién, on Panama's Caribbean coast, are more than 400 islands that make up the San Blas Archipelago. About 50 are inhabited and known as the Kuna Yala, homeland of the Kuna people. This tightly knit society did not move to the archipelago until the mid-1800s, but its folk have paid little heed to the trappings of the twenty-first century and continue to live much as their ancestors did hundreds of years ago. Villages of round, thatch-roofed huts have been built on islands with fresh water or close to the mainland where water is close by – it is the most valuable commodity here. Other islands are left uninhabited, but harvested regularly for their coconuts. Fishing, hunting and subsistence farming are practised, and the Kuna trade crabs and lobsters with the outside world. Men travel around the islands in dugouts, *ulu*, which can be paddled, sailed or nowadays have an outboard motor attached. The men also collect lobsters by free diving, going down 30 metres (90 feet) with one breath.

The Kuna's most eye-catching goods, though, are the famous *molas*, the colourful blouses worn by the women, not just for special occasions but for everyday wear. Each blouse is elaborately embroidered with panels (the *molas* that give the garment its name) depicting geometrical designs, plants and animals and traditional themes from Kuna culture, although nowadays pictures of cartoons on television, political images and other modern illustrations are beginning to creep in. It was partly the *mola*, or rather the government's attempt to dissuade Kuna women from wearing it and therefore abandon visible signs of their tradition and become more 'civilized', that led to a rebellion in 1925, after which the Kuna became largely autonomous, a status they retain to this day.

Further up the coast is Portobelo, named by Columbus on his fourth and last voyage to the New World, and the place where South America's gold left Central America for Europe. Drake was one of the English privateers who hijacked the Spanish treasure ships and tried to divert their contents to English coffers. Nearby Peñón de Drake or Drake's Rock is said to be the site where, after he died of dysentery, he was laid to rest in a lead coffin on a coral reef. Divers still search for the coffin, but it is now probably covered entirely by encrusting corals.

OPPOSITE The brown-throated or three-toed sloth *Bradypus variegates* has an unstable body temperature that can vary with its surroundings.

About 30 km (20 miles) to the southwest of Portobelo, on the other side of Madden or Alajuela Lake, is another Central American icon, the Panama Canal. Work on the canal started in 1881, and it opened in 1914, although the idea was suggested in Spanish treasure-ship times, as far back as 1524, with plans drawn up in 1529. The benefits of not having to round Cape Horn were appreciated even by the early European traders. A ship travelling from San Francisco to New York, for example, slices 12,600 km (7827 miles) off its journey.

On average 14,000 ships pass through the canal each year. The longest was the *San Juan Prospector*, 229 metres (751 feet), and the fastest passage was by the US Navy hydrofoil *Pegasus*, which crossed in 2 hours 41 minutes (the average is 8–10 hours). The slowest journey was made not by a boat, but by Richard Halliburton, who swam through in 1928. He also paid the lowest ever toll, just 36 US cents.

National parks are located right alongside the canal. Close to the east bank and just 25 km (15 miles) from Panama City, for example, is the Soberania, a tropical moist forest of cotton tree *Ceiba pentandra* and pink trumpet tree *Tabebuia rosea* known to be exceptionally rich in bird life. The world record for the number of species seen in a single day was set here in December 1992, when 525 were recorded in the traditional Christmas bird count along the famous (among birders, that is) Pipeline Road. Ticks might include the very rare crested eagle *Morphnus guianensis* and the beautiful red-lored parrot *Amazona autumnalis*.

Camino de Cruces National Park is also adjacent to the canal. Passing through it is the cobble road along which Peruvian gold was hauled on its way to Spain. The forest, streams, waterfalls and diverse fauna, including the Panamanian tamarin *Saguinus geoffroyi* and the three-toed sloth *Bradypus* spp., are much the same today as they were when 16th-century Spanish conquistador and explorer Vasco Núñez de Balboa and his companions walked here all that time ago. Today, these forests are important not only for protecting wildlife and ancient roads, but also for the very survival of the Panama Canal. If they were removed, the canal would soon silt up.

Offshore, on the east coast, is the Bocas del Toro Archipelago, claimed to have the best diving sites not only in the whole of Panama but also along the entire Caribbean coast of Central America. On the Isla Bastimentos, for example, is the Parque Nacional Marino where over 200 species of fish can be seen swimming among the corals; nurse sharks glide gently to a standstill on the sandy seabed;

four species of sea turtles come to nest; upside-down jellyfish *Cassiopea* spp. rest the wrong way up on the lagoon floor while catching plankton. Plankton, however, only makes up a very small proportion of this jellyfish's diet for algae living in its poisonous tentacles supply it with 90 per cent of its nutrients.

The islands themselves are varied – some with forests, others simple coral cays or thickly wooded mangrove islets. Mantled howler monkeys inhabit the forests. They are up and about at dawn howling loudly to announce that they are 'at home'. They are leaf-eaters, and as such require a small home range, so thrive on these islands. Each monkey consumes the equivalent of a human eating 32 lettuces a day. Leaves are particularly difficult to digest, so the howlers have a large stomach and the habit of resting during the what's left of the day.

The mangrove islets are also home to a unique race of three-toed sloth, one of the few places where sloths have adapted to living in this kind of vegetation. They even swim from one stand to another, the only sloths in the world to swim in seawater. They are also smaller than their cousins on the mainland, one island having especially small animals that are now considered a distinct species.

In fact, these islands are considered by evolutionary biologists to be the 'Galapágos of the Caribbean' for several islands have species that must have evolved from common, mainland ancestors; and none is more evident than the poison-dart frog *Dendrobates* spp., an example of evolution in action. Each island has its own species, sporting its own colour. There are yellow frogs with green spots, green frogs with yellow spots, red frogs with blue trousers, red frogs with blue spots and so on. Some are restricted to relatively small habitats. On Isla Bastimentos, for example, the local strawberry poison-dart frog *D. pumilio* – a tiny bright-red frog with black polka dots – is confined to a single beach on the North Shore that is aptly called Red Frog Beach.

Away from the shore, the dry forest is occupied by night or owl monkeys *Aotus* spp., the only truly nocturnal monkeys, a distinct race of golden-collared manakin *Manacus vitellinus*, one of a group of small and colourful forest birds known for their exuberant courtship dances and the tree-living eyelash viper *Bothriechis schlegelii*. Some individuals of this venomous species sport an eye-catching, neon-yellow body colour, but whatever their hue they all have a large scale above each eye, the 'eyelash' that gives them their common name. It is thought that this helps camouflage the snake by breaking up the outline of its head against the background of branches and leaves. The viper also has a prehensile tail, enabling it to snatch and swallow its prey while hanging upside down from a branch.

The pounding surf of the open ocean constantly batters the North Shore of Bastimentos, so conditions are always pretty rough. This means fewer tourists, making several beaches popular with nesting turtles, especially Playa Larga. The waves also attract another kind of marine 'creature' – the dedicated surfer; one of the biggest surf breaks in the world is right here. It is called Silverbacks, and the sport's aficionados compare these waves to Hawaii's monstrous Backdoor. It's a right-hand reef-bottom point break that barrels and can exceed 7 metres (20 feet) high, a size more usually associated with the famous West Peak, at Sunset Beach on Oahu's north shore.

Next door to Bastimentos is Isla Colón, and off its northern shore is Sway Cay or Isla de los Pájaros – Island of Birds. This small island has the red-billed tropicbird *Phaethon aethereus*, a spectacular creature that is white with a black trailing edge to its wings and a red bill, like many seabirds, but differs from the others in having trailing tail feathers that can be up to 1 metre (3 feet) long. It is also known as the bosun bird on account of a call reminiscent of a bosun's whistle. It feeds out at sea, plunge-diving to catch fish and squid, and then returns to its rocky ledge nest site, sometimes only a couple of metres above the level reached by the highest storm waves. About 30 pairs breed here, the only known nesting colony in the southwest Caribbean.

The wildlife superlatives continue in the next country to the north, Costa Rica. Take its plant life alone: it boasts 1700 species of orchids and 200 species of bromeliads, and they are to be found in protected areas ranging from the cloud forests of Monteverde and rainforests of Braulio Carrillo National Park to the seasonal dry forests of Santa Rosa National Park. Supporters of the dry forests claim that there are so many species of tree, with exotic names like poró, jacaranda, corteza and flame of the forest, that every one you pass is different from its neighbour. At the start of the dry season, the deciduous forest is awash with different-coloured blooms – red, white, purple, pink, orange and yellow – and at the start of the wet it happens all over again with a vivid array of completely different species.

Costa Rica's Caribbean east coast is hot, tropical and dominated by bananas, but many sections are protected as national parks or wildlife reserves. Each year from March till May, Manzanillo's 9-km (6-mile) long black-sand beach in the Gandoca-Manzanillo wildlife refuge near the border with Panama is host to four species of sea turtles, including the leatherback *Dermochelys coriacea*, while the nearby Gandoca lagoon and estuary have resident crocodiles, caiman, manatees, a host of waterbirds and the only stand of red mangroves in Costa Rica. Tarpon *Megalops atlanticus*, a popular gamefish, come here to spawn, and there is a large bank of mangrove mussels *Crassostrea rhizophorae*.

Although Manzanillo has a black sand beach, Gandoca's is white. Offshore, a pod of dolphins is often seen, while inland the forest, dominated by the 30-metre (100-foot) tall cativo or camibar tree *Prioria copaifera*, has 358 species of birds, including the chestnut-mandibled toucan *Ramphastos swainsonii* and ornate hawk-eagle *Spizaetus ornatus*.

Costa Rica's Caribbean east coast is hot, tropical and dominated by bananas, but many sections are protected as national parks or wildlife reserves

Just 45 km (30 miles) to the north is Cahuita National Park with a shoreline of white sand beaches and coral reefs, and a swamp that sits in the depression between a coral platform and the mainland. It is the proud possessor of an eighteenth-century wreck, thought to be a slave ship that went down near the mouth of the Perezoso (meaning 'slothful') River. Moray eels, sharks and barracudas now guard the ship, and three species of angelfish *Holocanthus* spp. and blue parrotfish *Scams coenileus* swarm over the reef that claimed it.

Coconut palms *Cocos nucifera* and holillo palms *Raphia taedigera* grow right to the edge of the sea, leaning towards the ocean in a picture-postcard landscape, but it was all nearly destroyed in an earthquake in April 1991, when trees were demolished and parts of the coral reef were pushed up by 3 metres (10 feet), bringing them above the sea's surface. Nevertheless, the surviving reef is beginning to heal, and 35 species of corals now grow here. The forests and mangroves are recovering too, with green ibis *Mesembrinibis cayennensis* and northern boat-billed herons *Cochlearis cochlearis* in residence, and green-and-rufous kingfishers *Chloroceryle inda* hiding shyly behind the screen of green leaves. Land crabs are all over the place, with four easy-to-spot species: the black land crab *Gecarcinus lateralis*, which is actually dark red, the white land crab *Cardisoma guanhumi*, which is bluish-white, the mangrove crab *Ucides cordatus*, which is red, and the land or tree hermit crab *Coenobita clypeatus*, which has red-orange legs and a darker maroon-red body. They become active from sunset to about eight o'clock in the evening and then gradually slow down until noon the following day when they stop altogether to avoid desiccation.

On Costa Rica's northwest coast, the Tortuguero National Park is known worldwide for its turtles. It came to prominence in the 1950s when American biologist and writer Archie Carr formed the Brotherhood of the Green Turtle (now the Caribbean Conservation Corps), which worked with the Costa Rican government to protect an area that is now the most important site in the Caribbean for nesting green turtles *Chelonia mydas*. Over 30,000 come here between June and October each year, the greatest numbers arriving in September. The females haul themselves laboriously across the brown sand, and each deposits more than a hundred eggs in a pit she has dug for herself at the top of the beach. And they do not come just once. Each female may haul out two to six times at 10–14 day intervals and will then wait for two or three years before nesting again.

The hatchlings appear about 60 days later, all the baby turtles from a single nest emerging synchronously at night and heading unerringly towards the sea. Coatis *Nasua* spp. sometimes dig up the nests and eat the developing eggs; black vultures *Coragyps atratus* scavenge on the leftovers; and, if the hatchlings survive that, they are attacked on their way to the sea by iguanas, ghost crabs and frigatebirds. Should they reach the water, then the sharks are waiting for them. How any survive at all is a miracle, but survive they do. After a 'missing' year at sea

ABOVE The water hyacinth is the world's most serious acquatic weed, notorious for blocking waterways and causing lakes to stagnate.

(so called because they simply disappear and until recently nobody knew where they went), during which the little hatchlings drift with the ocean currents and hide under floating weed, they grow to a respectable egg-bearing size, and the females return several years later to deposit their own eggs in the sands of Tortuguero. The males wait offshore to mate with the receptive females.

The beach here is 35 km (20 miles) long, of which 22 km (14 miles) is protected. Behind it is a narrow lagoon that stretches its full length, open to the sea at one end, fed by a river at the other, and fringed by coconut palms. Inland is a coastal rainforest, crossed by a mosaic of river channels and adjacent to swamps fed by streams from low-lying volcanic hills, such as Coronel, Caño Moreno and the Sierpe Peaks. The swamp forest is dominated by holillo palms, some reaching 40 metres (130 feet) tall, together with wild tamarind *Pentaclethra macroloba*, crabwood *Carapa guianensis* and cativo. Open-water channels are often blocked by water hyacinth *Eichhornia crassipes*.

Over 300 bird species have been recorded here, such as the endangered green macaw *Ara ambigua* and keel-billed toucan *Ramphastos sulfuratus* and the more common swallow-tailed hawk *Falco furcatus*. Reptiles include crocodiles, iguanas and basilisk lizards *Basiliscus* spp., as well as two other species of sea turtles –

leatherbacks that come in April and May and hawksbills *Eretmochelys imbricata* that appear in July. Sixty species of frogs and toads are recognized, including the glass frog *Centrolenella valerioi* whose males are so diligent that they will guard their egg masses on the undersides of leaves overhanging a stream even in the daytime, when it is dangerous for little frogs to be out and about, and another colour form of the strawberry poison-dart frog. Unlike the spotted ones at Bocas del Toro, these have a bright-red body and purple legs, and they show even more devoted parental care than the glass frogs.

The clutch of up to five eggs is protected by mucus and is visited regularly by both parents to ensure it remains moist. When the time comes for the tadpoles to hatch, one of the parents wades into the mass, freeing the tadpoles, who slither on to its back and are anchored there for the short journey to water in a bromeliad reservoir or a puddle. Each tadpole is placed in a separate body of water in order to prevent cannibalism among the siblings, and after three weeks the little froglets return to the forest.

There are also 60 mammal species here, including endangered jaguars that have been seen taking turtles on the beach, tapirs that forage there, and manatees that were thought to be extinct here until a small population was found in a few remote lagoons. Once upon a time the manatees were killed for their flesh and tough hides, but currently increased tourist-boat traffic and agrochemicals and sediments washed into the water from banana plantations are the major threats. There are thought to be no more than a hundred manatees left in Tortuguero and the Barra del Colorado National Wildlife Refuge, both part of the Tortuguero Conservation Area that runs to the border with Nicaragua.

Nicaragua is another biologically diverse country, due to a low human-population density and the presence of Central America's largest expanse of forest, much of it protected by 78 national parks and wildlife reserves and containing trees such as Nicaragua's national tree, the white-flowered sacuanjoche or frangipani *Plumeria alba*, and such brightly coloured birds as scarlet macaws *Ara*

OPPOSITE A strawberry poison frog on vegetation on Panama's Caribbean coast.

macao, green parakeets *Aratinga holochlora* and the country's national bird, the beautiful guardabarranco or turquoise-browed motmot *Eumomota superciliosa*.

OPPOSITE One of the few strongholds of the Great green macaws is the Darién Biosphere Reserve in Panama.

The Caribbean coast of Nicaragua is a wide, flat coastal plain about 540 km (335 miles) long and covered by forest. Many rivers and streams drain to the east across the plain, including the Río San Juan on the Costa Rica border, the outflow from Lake Nicaragua that is just 17 km (10 miles) from the Pacific Ocean yet flows into the Caribbean. The river and lake are famous for their 'freshwater' sharks – in reality bull sharks *Carcharhinus leucas*, a marine species whose females can adapt to fresh water in order to enter rivers and lakes to drop their pups, a place safe from other sharks, including males of their own species.

This coastal strip, of brackish lagoons and forested sand spits, is about 60 km (40 miles) wide and extends northwards from the San Juan to northeast Honduras, a wilderness area known as La Mosquitia or Mosquito Coast, the biggest area of flat wetlands in the Americas and an important overwintering site for bird migrants. It is named not after the pesky insect, although there are plenty of these about, as well as swarms of sand flies, but after the Mosquito Amerindians, one of several tribes that live here. The area consists mainly of impenetrable jungle, through which there are no roads. Transport is mostly on rivers by dugout canoe. There are several lagoons, the largest being Laguna de Perlas, where tourists have yet to invade and fishermen still catch fish, lobsters and shrimps.

Offshore are several groups of islands, including the Islas del Maíz or Corn Islands, about 75 km (45 miles) from the mainland. Some, like the Pearl Cays, about 27 km (17 miles) from Big Corn Island, are barely explored, but just over 1 km (⅔ mile) to its southeast is the wreck of a Spanish galleon in 20 metres (70 feet) of water, one of many that went down on this stretch of coast, the result of raids by buccaneers. It is thought that some might still contain treasure.

Further to the north and five hours' boat ride from the mainland is Cayos Miskitos, an archipelago of 85 islands fringed by mangroves and surrounded by coral reefs, once a haven for pirates and turtle hunters but now the province of lobster fishermen who have discovered a new sport. They collect 'white treasure', the discarded packages of cocaine dumped by traffickers pursued by coastguards.

All these islands have lost much of their original vegetation because livestock grazed on it – especially pigs and goats who were brought here by the pirates and buccaneers so that fresh meat was available on demand. The trees were also cut for lumber, especially the mahogany *Swietenia mahogany,* or replaced by introduced coconut palms. Nevertheless, isolated stands of native moist-forest trees such as fustic and cedar *Cedrela odorata* are still evident.

The area of Mosquito Coast consists mainly of impenetrable jungle, through which there are no roads. Transport is mostly on rivers by dugout canoe

A handful of offshore islands here are governed not by Nicaragua, but by Colombia, 620 km (385 m) away to the southeast. The San Andrés Archipelago is a scattering of coral islands that sit on ancient volcanic cones. San Andrés itself is the largest and home to two endemic land birds, the San Andrés mockingbird *Mimus magnirostris* and the San Andrés vireo *Vireo caribaeus*, as well as an endemic lizard *Anolis concolor* and a snake *Coniophanes andresensis*. The fauna includes two species of land crabs – the black crab *Gecarcinus ruricola* and shankey *G. lateralis* – both of which embark on mass migrations to the sea to deposit their eggs. The event takes place for about two weeks in May and June, when each female contributes about 40,000 eggs to the living soup along the shore. One of the smaller islands, Queena, has a semi-circular coral reef that is 60 km (40 miles) long and up to 20 km (12 miles) wide, the most extensive in the archipelago and the least studied. It is thought that commercially valuable lobsters and queen conch are still here in appreciable numbers.

OPPOSITE The jabiru is a large stork whose name comes from the Tupi-Guaraní language of South America, meaning 'swollen throat'.

Back on the mainland, the Mosquitia coast of Nicaragua is separated from that of Honduras by the 800-km (500-mile) long Coco (formerly Segovia) River, Central America's longest river, which rises about 80 km (50 miles) from the Pacific coast but has its outflow at Cabo Gracias a Dios on the Caribbean side. It is a main highway through the Mosquitia region, but it is not always benign. In October 1998, when category 5 Hurricane Mitch hit the region, the river flooded vast areas, making hundreds of thousands of people homeless and killing tens of thousands.

Behind the shoreline of the Honduras section of La Mosquitia, another major waterway has lent its name to the Río Plátano Biosphere Reserve, home to 5000 Mosquito and Paya Amerindians and the world's first World Heritage site. About a quarter of the reserve is coastal plain, while three-quarters is mountainous, topped by Punta Piedra at 1326 metres (4350 feet). There are stunning 150-metre (500-foot) high waterfalls and equally high natural rock formations like the granite pinnacle at Pico Dama. Ridge tops have elfin forest, and the rest is cloaked in very humid tropical forest, broad-leafed gallery forest and, at the coast, mangrove swamps, sedge prairie and coastal pine savannah. It is said that there are more trees per hectare here than in the Amazon rainforest. All the usual Central American plants and animals are present, notable bird species including the king vulture *Sarcoramphus papa* and three strikingly colourful species of macaw – scarlet, green and military *Ara militaris*. Estuary waters have the cuyumel or bobo mullet *Joturus pichardi*, an important food fish that lives and spawns in brackish waters, and feeds by rasping algae from rocks with its fleshy lips.

The park is the possible location of yet another lost city – Ciudad Blanca, the legendary 'white city'. It is supposed to be close to white sand deposits or carved from white rock, hence its name, but nobody knows where or what it is. It was first described by Hernán Cortés in 1526, when searching Honduras for the fabled town of Hueitapalan – 'Old Land of Red Earth' – the Central American equivalent of

El Dorado, where it is said that those of noble birth ate on plates of solid gold. Hueitapalan has also been linked in mythology with the birthplace of the Aztec god Quetzalcoatl.

Over the years, various aircraft pilots and hunters in the jungle claim to have seen the ruins of a forgotten city, and an expedition led by Theodore Morde in 1939 supposedly found a lost city that was described in his book *City of the Monkey God*. Its location in this vast and uncharted area, however, was not revealed, and Morde was run over by a car in London before he could divulge any information to the sponsors who were backing his next expedition. Since then, New Orleans engineer D. H. Williams is alleged to have visited the place while prospecting for petroleum, but he too kept the location secret lest looters ransack the site. Other expeditions have failed to find it.

On the north coast, the large mangrove estuary of the Cuero and Salado rivers is one of the most important manatee habitats in the western Caribbean. On the back of fruit-tree cultivation on the surrounding land, the privately administered (by an NGO) Cuero y Salado Wildlife Refuge was created to protect not only these aquatic mammals but also birds like the jabiru *Jabiru mycteria*. Water is provided by 15 rivers that originate in the cloud forest Pico Bonito National Park to the south, and access to the park is via a tram, a leftover from its United Fruit plantations days.

Like most of the other Central American countries, Honduras has its own offshore paradise islands. Between Honduras and the Cayman Islands, about 160 km (100 miles) north of the mainland, are the two Islas del Cisne or Swan Islands. Columbus found them in 1502, and named them Islas Santa Ana, but they became a haunt of pirates and by 1775 they were appearing on maps with their current name. Surrounded by sheer cliffs, rough seas and protected by razor-sharp corals, the islands, with their 1400-metre (4600-foot) runway, have been a

guano-mining centre, a US weather station, a CIA transmitting station at the time of the Bay of Pigs invasion, a Honduran military outpost and a potential holiday resort. In October 1979, they were the site of a tragedy when a southward-migrating flock of chimney swifts *Chaetura pelagica* put down on the islands and stayed for a week. Hundreds died of starvation, until none was left, yet resident vitelline warblers *Dendroica vitellina*, smooth-billed ani *Crotophaga ani* and white-crowned pigeons *Columba leucocephala* were unaffected. The die-off is a mystery.

Earlier in the decade, in December 1974, however, there had been a happier event when Spencer Bennett and Randolph Moore, of the US National Oceanic and Atmospheric Administration, were based at the weather station on Great Swan island. During a storm, a Honduran fishing vessel was sinking and the weather-station crew picked up the SOS. They launched two small boats and, in an operation reminiscent of Grace Darling's epic rescue in tempestuous seas, brought home the entire crew of 19 unharmed. Both men received the US Department of Commerce Gold Medal.

ABOVE The double-crested basilisk *Basilicus plumifrous* is known for its ability to run across water, earning its nickname – 'Jesus Christ lizard'.

OPPOSITE Mangroves seem to almost walk into the water on Roatan Island, Honduras.

About 25 km (40 miles) to the north of mainland Honduras are the Islas de la Bahía or Bay Islands, a 50-km (30-mile) long island arc that sits on the Bonacca Ridge, an underwater extension of the Sierra de Omoa mountains. The three main islands – Utila, Roatán and Guanaja (Bonacca) – together with 65 smaller cays are covered with pine and oak savannah and dry tropical forest, fringed with mangroves and surrounded by coral reefs. The pine forests on Bonacca inspired Columbus to call it the Isle of Pines. He was here on 30 July 1502, and it was from these islands that he first had sight of the mainland of Central America, where he searched in vain for the great cities of Asia.

Today, as then, there are about 40 species of reptiles flourishing in the relatively dry conditions on these islands. In the Carambola Botanical Reserve on Roatán, for example, a sheer cliff known as Iguana Wall is a breeding site for iguanas. On Utila, the Utila spiny-tailed iguana *Ctenosaura bakeri* is unusual in that it lives exclusively in the 12 sq km (5 sq miles) of mangroves that protect the island's coral reefs, only venturing on to the beach to deposit its eggs. It is the only iguana to live its entire life in the mangroves, where it is known locally as 'the swamper'. It feeds not only on mangrove flowers but also on seafood. At low tide, the iguanas invade the mudflats around the mangrove roots and catch mud crabs.

In fact, Utila is an extraordinary place. It is a small, flat island covering no more than 42 sq km (16 sq miles) yet it is packed with a variety of wildlife, including 33 species of reptiles and amphibians.

Another Bay Islands endemic is the basilisk, a medium-sized character that has the endearing habit when in a hurry of rising on to its two back legs and running along with its fin-like crests expanded. Snakes are represented by 13 species, including the venomous coral snake *Micrurus nigrocinctus*, known also as the 20-minute snake, for it is said that a victim will succumb to its venom after 20 minutes.

ABOVE LEFT
A Central American coral snake Micrurus nigrocinctus rests on the forest floor in Selva Verde, Costa Rica.

Fortunately, its fangs are short and fixed so far back in the jaw that injecting venom after a single lunge is difficult. Instead, it attacks repeatedly, eventually grabbing hold of the skin, hanging on and pumping in its poison. This snake is a handsome beast, its colourful stripes earning it the alternative name 'harlequin snake'.

The jewels in the Bay Islands' crown, though, are the magnificent coral reefs and cays. It is said that the islands are so replete with marine life that 96 per cent of all known species of Caribbean marine life is found here. From February to May, whale sharks *Rhincodon typus* appear. They feed on the small baitfish that also attracts bonitos *Sarda sarda*, the two species feeding side by side in a 'fish boil'. The whale sharks sometimes hang vertically in the water and actively 'suck' in the small fish, the only shark known to feed in this way. In April and May, the ocean currents bring in swarms of tiny thimble jellyfish *Linuche unguiculata*, and upwellings of nutrients from the seabed fatten up the plankton and small fish that feed huge shoals of jacks. The larger fish, in turn, are food for such unusual visitors as marauding schools of false killers whales *Pseaodorca crassidens* and even pods of killer whales or orcas *Orcinus orca*.

ABOVE The bull shark is one of the world's most dangerous sharks, but even this species can be won over with the offer of a free handout of fish.

Dive sites on Roatán have evocative names such as Eel's Garden, Dolphin Cave, Jagged Edge and Valley of the Kings, and they feature such underwater stars as Atlantic spinner *Stenella longirostris* and bottlenose dolphins *Tursiops truncatus*, hammerhead *Sphyrna* spp. and nurse sharks, and a colourful assortment of reef fishes. West End Wall has pillar corals and azure vase sponges *Callyspongia plicifera*, large barrel sponges *Xestospongia muta*, schools of horse-eyed jacks *Caranx latus* and passing eagle rays *Myliobatis* spp., while the Enchanted Forest has creole wrasse *Clepticus parrae* and a seemingly endless variety of groupers of the family Serranidae.

Two hours' boat ride from Roatán is Barbareta, a palm-lined, privately owned island with virgin rainforest, white beaches and coral reefs. Southern stingrays *Dasyatis americana* ply the shallows in Barbareta Marine National Park, but the main underwater attraction here is the Morat or Barbareta Wall, a series of overhangs and crevices, in which hide gigantic spotted spiny lobsters *Panulirus guttatus* and the sheep crab *Loxorhynchus crispatus*, one of the decorator crabs, a group of species that cover themselves with seaweed in order to blend in with their background. Deeper parts of the wall have black corals, gorgonians and a profusion of sponges, while the shallower shelves have elkhorn *Acropora palmata*, staghorn *A. cervicornis* and brain corals *Diploria* spp., and barrel sponges. Electric blue and purple Pederson's shrimp *Periclimenes pedersoni* are everywhere, plying their cleaning services to a clientele of groupers and snappers. Schools of vertical-striped, plankton-eating sergeant-majors *Abudefduf saxatilis*, schoolmasters *Lutjanus apodus* – a type of snapper found near elkhorn corals – and brilliantly coloured blue tang *Paracanthurus coeruleus* (Nemo's best friend in the feature film *Finding Nemo*) swirl this way and that across the reef.

The northern barrier reef off Guanaja has pinnacles and volcanic caves, steep drop-offs and walls, blind tunnels and the painful fire coral *Millepora* spp. Fire coral is not a true coral, but more closely related to sea anemones and jellyfish. It has a bright yellow-green or brown skeleton and is easily mistaken for seaweed but, should you brush against it, its hard and jagged exterior abrades the skin and its sting cells shoot venomous darts into the raw flesh. Five to ten minutes later you feel a burning sensation; in severe cases it can cause nausea and vomiting, and the symptoms can last long enough to ruin a vacation. Fire coral, which can be in all manner of shapes depending on the species, is distinguished from the true corals by the white edging across the tips of the branches.

The reefs of the Bay Islands are truly impressive, with or without fire corals, but they are just a taster for the natural spectacle a little further to the north, for the Bay Islands are part of the Meso-America Reef that hugs the Caribbean coast of Honduras, Guatemala, Belize and Mexico.

Off the coast of Belize, the reef becomes the second-largest barrier reef in the world, beaten to the record books only by the Great Barrier Reef of Australia. It is nearly 260 km (160 miles) long, stretching from the country's northern border, where it is just a kilometre (⅔ mile) offshore, to Sapodilla Cays to the northeast of Guatemala, which are 40 km (25 miles) from the coast. It is also one of the world's most diverse coral-reef systems, with the main barrier growing along the edge of the continental shelf, fringing reefs along the mainland coast, patch reefs, which can be simple coral clumps or sprawling reefs and feature strongly at Southwater Caye and faroes, which are steep-sided, atoll-like rings of coral with a lagoon at the centre, the finest example to be found at Laughing Bird Caye.

There are also true coral atolls, a feature generally associated with underwater volcanic peaks and more usually seen in the Pacific Ocean. There are four in the Atlantic region, three of them off Belize (the fourth, Banco Chinchorro, is further north, off Mexico). The Belize atolls are located on non-volcanic submarine ridges, the most southerly being Glover's Reef, named after the English pirate John Glover who plundered Spanish galleons from his base here. Much of the reef is unexplored and unexploited, a diving heaven.

Some of the reefs have large islands with coastal wetlands, lagoons and sea-grass meadows fringed by mangroves. The entire formation led one famous coral-reef expert, none other than the naturalist Charles Darwin, to remark that this was 'the most remarkable reef in the West Indies'.

Biological diversity is phenomenal. A single reef is home to a minimum of a thousand marine species, and on the Belize barrier reef it is estimated that 90 per cent of its animals, mainly invertebrates but including new species of fish, have yet to be discovered. In the past couple of years, for example, a 'biodiversity hotspot' was revealed at the mangrove-lined Pelican Cayes, where not only do unusual communities of sponges and corals exist but also over 40 species of sea squirts occur in a single lagoon.

The reef is also steeped in history. The Turneffe Islands, the largest of Belize's coral atolls with over 200 cayes, was the centre of a flourishing sponge fishery in the 1930s, and its mangrove-lined channels and tiny, uninhabited islets were the haunts of pirates said to have captured women from Bacalar, over the border in Mexico, and brought them here. Today, the coral reef is home to the rare and primitive-looking white-spotted toadfish *Sanopus astrifer*, which is endemic to Belize and only comes out at night, and the mangroves provide a covert nursery for the rare American crocodile *Crocodylus acutus*. The local egrets and herons, however, are in on the secret and swallow the crocodile hatchlings whole.

St George's Caye was also frequented by pirates and buccaneers, but it became the first capital of the British settlement in the region in 1650, and was soon a target for the Spanish. It all came to a head on 10 September 1798, when a Spanish war fleet of 32 ships carrying 2000 troops and 500 crewmen headed for the island. The British were hopelessly outnumbered: the Baymen of Belize, as they were known, had one large sloop, HMS *Merlin*, and a couple of smaller boats. Since the original landing 150 years earlier, many people had moved to the mainland, but at the first sound of gunfire both settlers and troops acquired every little fishing smack, rowing boat, dory, canoe or pitpan they could lay their hands on, together with every gun they could muster, and went to help the Baymen. Miraculously, they won the day, and without loss of life. The defeated Spanish never harassed the settlement again.

On the Belize barrier reef it is estimated that 90 per cent of its animals, have yet to be discovered

The major reef most distant from the mainland is Lighthouse Reef, an atoll with numerous patch reefs and a sandy seabed

LEFT The almost perfectly circular Great Blue Hole, at Lighthouse Atoll on the Belize Barrier Reef, is a collapsed and submerged cavern about 0.4 km (¼ mile) across.

ABOVE One of the many wrecks that litter the Caribbean, divers surprise a shoal of squirrel fish *Holcetrus* spp.

The major reef most distant from the mainland is Lighthouse Reef, an atoll with numerous patch reefs and a sandy seabed. There are several cayes – Sandbore, White Pelican, Long and Hat cayes – but one of the most inviting is Half Moon Caye, at the atoll's southern end. It has a colony of nesting red-footed boobies *Sula sula* and magnificent frigatebirds *Fregata magnificens* on the land, and the amazing Half Moon Caye Wall underwater, a sheer drop of 900 metres (3000 feet) with surfaces covered with barrel and tube sponges, sea fans and gorgonians. Shallow-water areas have sand flats with garden eels *Heteroconger longissimus*. These small, eel-shaped fish keep their tails in burrows, and when danger threatens they withdraw, as if they had never been there at all.

The reef has several wrecks perched high and dry on the coral, including one that was once used by the RAF for target practice. The pilots were so accurate that the ship became known as 'Broken in Two Wreck', but was later renamed 'Harrier Wreck' after the aircraft used to destroy it.

The lagoon at Lighthouse Reef has shallow-water pools and caves with what have been described as 'very red' lobsters, and in the early 1970s divers watched what they called a 'conch walk': hundreds of adult conch, each about 3 metres (10 feet) from its neighbour, all slithered along in a generally southwesterly direction at dusk. No one knows where they were going or why – another Caribbean mystery.

The most spectacular feature here, though, is the Blue Hole, an almost perfectly circular limestone sinkhole, 300 metres (1000 feet) across and 125 metres (400 feet) deep. It was once the entrance to an underground cave system formed during the Ice Age when the sea level was considerably lower than it is now. After the ice caps melted, the seas rose and flooded the labyrinth of caves and passageways, some of which still have stalagmites and stalactites on their floor and ceiling. Today, the rim is lined with sponges and corals rather than ferns and flowering plants, and the shallower waters inhabited by barracudas instead of bats.

Deeper down, below 18 metres (60 feet), less light penetrates, and the walls are bare. The only companions here are the occasional hammerhead and gangs of menacing bull sharks, a species that has recently moved in. Caverns contain the skeletal remains of sea turtles that found their way in but failed to find a way out. Dive boats reach the centre of the dark blue hole following a route pioneered by French underwater explorer Jacques-Yves Cousteau's ship, the *Calypso,* in the early 1970s (with the help of a few carefully placed explosives, if local diving folklore is to be believed).

According to visiting divers, the hole also once had its very own sea serpent. An expedition in the early 1970s claimed to have seen a strange creature about 6 metres (20 feet) long, with red

They might be the biggest fish in the sea, but they feed on the smallest

OPPOSITE The whale shark is a frequent visitor to Caribbean reefs where it synchronizes its arrival with spawning reef fish.

eyes, a translucent eel-like shape and long, flowing fins. Nitrogen narcosis, a condition similar to intoxication by alcohol and the result of diving at depth for too long, might explain the 'monster', but there is also a more down-to-earth suggestion. It could have been a wayward oarfish or king-of-herrings *Regalecus glesne*, a long, silvery, eel-like fish that grows up to 8 metres (25 feet) long; it is the world's longest bony fish, but one that more usually lives in the open ocean.

Another open-ocean giant is a more frequent guest. In the south-central part of Belize, about 40 km (25 miles) east of the Placencia Peninsula and Lagoon, is a coral-reef promontory known as Gladden Split, and this area of the barrier reef hosts an extraordinary event involving the world's largest fish, the whale shark *Rhincodon typus*, known locally as Sapodilla Tom. Growing up to 20 metres (65 feet) long, these sharks can only be described as 'enormous'. They dwarf divers who swim with them, but they are harmless. They might be the biggest fish in the sea, but they feed on the smallest. They are filter feeders, generally sifting jellyfish *Linuche unguiculata* and copepods from the surface waters, but each year their arrival at Gladden Split coincides with the spawning ritual of cubera *Lutjanus cyanopterus* and dog snappers *L. jocu*. These are normally solitary reef-dwellers, feeding on other fish, crabs and spiny lobsters, but in the spring, at the time of the full moon from March to June, they abandon their usual haunts and go a-courting.

The sharks seem to know precisely when to come, although early birds have been known to wait for up to 14 hours for the fish to spawn. When the event gets under way, enormous aggregations of literally thousands of snappers spiral up from a depth of 30 metres (100 feet) to within 1.5 metres (5 feet) of the surface. Every now and again, groups break off and in a fast-swimming frenzy a few females and a lot of males scatter eggs and sperm into the water. The whale sharks sweep in to gather up the spawn. Curiously, the sharks congregate only at Gladden Split, ignoring similar spawning events at 13 other known sites When the Gladden glut is over, they travel hundreds of kilometres along the reef edge, diving to depths of more than 1000 metres (3300 feet) where the pressure is enormous and water temperature is less than 5 °C (40 °F); they are found as far afield as Honduras and Mexico in search of more widely dispersed food.

Ambergris Caye, in the most northerly part of the barrier reef, is Belize's most accessible reef area. This long, thin island, the Isla Bonita or 'beautiful island' featured in the Madonna song of the same name, consists mainly of mangrove forest and coral reefs. It is a popular site for divers, with hundreds of dive sites such as Shark-Ray Alley, where nurse sharks and stingrays congregate to be fed and petted at a place where fishermen once cleaned their catch and discarded the waste into the sea. Not far away, indigo hamlets *Hypoplectrus indigo* engage in a curious and complex courtship ritual, during which a pair of fish swaps sexual roles, playing both male and female, for they are hermaphrodites.

Ambergris Caye is a geological extension of Mexico's Yucatán Peninsula, a gigantic promontory of porous limestone rock, from which it is separated by a narrow cut, no more than the width of a boat. On the other side, the first things that strikes a visitor are the absence of rivers or streams and the presence of numerous waterholes that the locals call *cenotes*, meaning 'wells'. The ancient Maya, who once had a great civilization in this region, thought of *cenotes* as the entrances to another world, a mysterious and spiritual underworld inhabited by fearsome gods; and in a way they were right, for underneath the surface is an extraordinary labyrinth of uncharted, water-filled caves and passageways that cave divers refer to as 'inner space'. These cave systems are so extensive that explorers have discovered one of the longest caves in the world, over 133 km (83 miles) long. They are underground rivers, the main source of water for both humans and wildlife alike, and they drain from the Yucatán to the Caribbean Sea.

The pools visible on the surface are dotted throughout the jungle and are havens for birds, like the hooded warbler *Wilsonia citrina*, which travels from as far away as Newfoundland to spend the winter here, and turquoise-browed motmot; they are also watering holes for peccaries, deer and tapirs that visit under cover of darkness, wary of jaguars in the tightly packed forest all around. The trees harbour black howler monkeys *Alouatta pigra,* which are such sloppy eaters that coatis are able to follow them about and feed on the falling scraps. In the crystal-clear water, pierced by shafts of sunlight, are fish such as sailfin mollies *Poecilia velifera*, the males busily keeping their harems together and chasing off rival males, and Mexican tetras *Astyanax mexicanus*, which have learned to follow the lights of divers to feed deep inside the caves, for even in the darkness there is life.

Some 30 known species lead their entire lives in the dark waters, many without pigment or eyes. The remipede *Speleonectes tulumensis*, a primitive form of crustacean that was until the 1980s known only from fossils, occupies estuary-like waters close to the sea and hunts down cave shrimps and isopods. The blind cavefish *Ogilbia pearsei* swims in fresh water among the roots of forest trees that have pushed down through the cave roof to tap the reservoir below. At one time, the caves must have been dry, for there are stalagmites and stalactites and fossils of ancient animals, such as the primitive elephant *Gomphotherium*, which must have lived on the surface. They are filled too with Maya artifacts and even human skeletons, evidence that they were central to the Maya culture.

Most of the *cenotes* are sprinkled randomly over the surface of Yucatán, but in a great semi-circular arc, about 165 km (100 miles) across, is a series of especially deep 'wells', some with a dark cloud-like hydrogen-sulphide layer at their bottom. Geological research has shown that the circle is completed on the floor of the Caribbean Sea. It marks the spot where a huge extraterrestrial body slammed into our planet, and it is thought that the impact was responsible for the mass extinction of 75 per cent of the plant and animal species living on the Earth, including the final demise of the dinosaurs, about 65 million years ago.

7

OUTER CARIBBEAN

PREVIOUS PAGES
An aerial view of Exuma Cays on the Grand Bahama Bank.

OUTSIDE THE inner circle of countries directly bordering the Caribbean Sea, there are groups of islands and cays with climatic, biological and cultural influences so close to those of the true Caribbean that it would be remiss not to include them. They are to be found in the northeast of the region, stretching from the Florida Keys, which are, of course, part of the USA, to the British-owned Turks and Caicos Islands in the southeast. But the biggest archipelago by far is that of the Commonwealth of the Bahamas that runs between the two. It was named not by Columbus but in 1513 by Spanish explorer and soldier Juan Ponce de León, during his quest for the 'fountain of youth', a legendary spring that gave people eternal life. He failed to find the spring, but he described the area around the Bahamas as *baja mar*, meaning 'shallow sea', a reference to the coral reefs and massive sandbanks that are sculpted here by the ocean currents.

Geographically, the Bahamas lie in a northwest-southeast orientation separating Cuba from the might of the Atlantic Ocean. They extend about 1400 km (870 miles) from Florida in the north to Hispaniola in the south, but their land area is minute. It totals no more than 14,000 sq km (5400 sq miles), but it is spread over 260,000 sq km (100,000 sq miles), an area of sea almost twice the size of Spain. Most of it is sandbank and coral reef; in fact, about 5 per cent of the world's coral reefs are here, and it all sits on a rocky platform, known as the Great Bahama Bank in the south and the Little Bahama Bank in the north. The banks were formed millions of years ago when the calcareous remains of marine organisms settled in layers on the ocean floor. These layers were continually buried and compressed to form massive blocks of oolite limestone that is currently about 8 km (5 miles) thick. Over 700 relatively flat islands and 2400 smaller cays sit on the platform, of which about 30 are inhabited and many are just a metre or two (a few feet) above the sea. The highest point is Mount Alvernia on Cat Island, its summit just 63 metres (207 feet) above sea level.

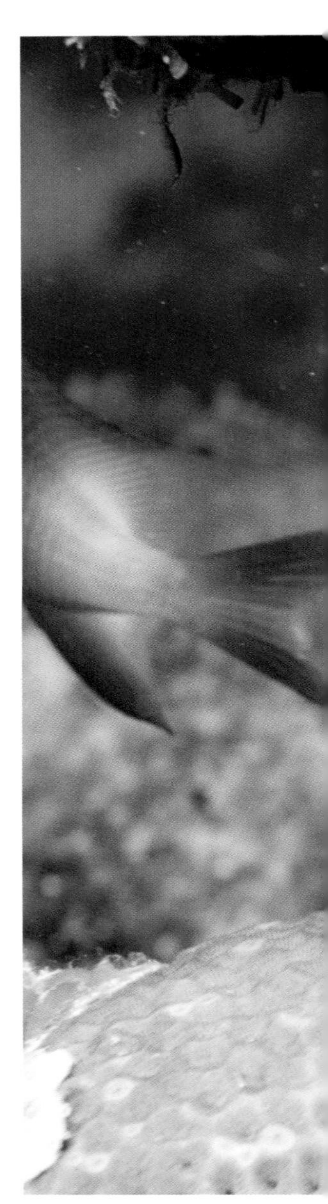

Thousands of years ago during the Ice Age, when the sea level was 120 metres (400 feet) lower than it is today, the Bahama banks were dry land and eroded like any other limestone scenery. Today, those features are under water. Deep-water channels such as the Tongue of the Ocean east of Andros Island are really ancient river valleys carved by torrents that flowed down to a shoreline some distance from the one we see today. The islands themselves are honeycombed with potholes, caves and crevices, the result of erosion by rainwater.

Though close to the North American mainland, the vegetation of the Bahamas has its origins in Cuba and Hispaniola. Mangroves, with their floating, salt-tolerant seeds, were probably first to arrive, followed by plants the seeds of which were brought by birds. Interestingly, of the 1200 species recorded on the islands today, the vast majority produce fruits and seeds that are eaten and distributed by birds. The pigeon plum *Coccoloba* spp., for example, bears a small, black fruit that satisfies resident birds in summer, and in late winter visiting

North American migrants feast on the dark, red fruit of the satin leaf tree *Chrysophyllum oliviforme*.

The islands are not lush – where natural vegetation prevails it is mostly dry forest and scrub, alongside wet thickets known as coppices. Nonetheless, there is a variety of trees, but the expectation of swathes of palm trees, the quintessential tropical-island tree cover, is replaced with a reality of introduced casuarinas or

BELOW Rock beautiful *Holocanthus tricolor*.

Australian pine *Casuarina equisetifolia*, which some consider an invasive species to be removed at the earliest opportunity. One of the natives is the giant poisonwood *Metopium toxiferum*, which contains the same allergenic oils as poison ivy.

Flowers are everywhere, from the red blooms of the bull vine *Cissus intermedia* to the salt-tolerant beauties of the shoreline, like the untidy white flowers of the spider lily *Crinum zeylancium*. Watch out for unpleasant surprises, though. The Bahamian haulback *Mimosa bahamensis*, an endemic mimosa has lovely pink flowers, but hidden underneath are long, sharp thorns.

The most arresting natural features in the Bahamas must be the blue holes, similar geologically to the *cenotes* of the Yucatán (see chapter 6). There are over 200 to found inland and another 50 accessible in the sea. Each is an ancient sinkhole, a giant cavern the roof of which fell in to leave a vertical shaft that is now the grand entrance to an intricate cave complex. Today, the entire system is under water, some holes containing fresh water, some salt and others both. Some 69 species of troglobitic animals have been observed in the caves, including sponges, copepods, opossum shrimps, remipedes, worms and the Bahamas blind cavefish or New Providence cusk-eel *Lucifuga spelaeotes*, one of the world's strangest and most endangered freshwater fish. It belongs to a family of deep-water marine fish that was isolated in these sinkholes when the sea level changed. It adapted not only to fresh water but also to living in the dark. It is blind and without pigment, about 10–12 cm (4–5 inches) long, and survives only in the Bahamas blue holes, living in the 'lens' of fresh water that sits on the heavier salt water below. Its nearest relatives are the blind cavefish of Cuba and the Galápagos Islands, members of the same genus. The main threats to its survival come from people who dump garbage and even sewage into the inland caves, and from divers who stir up the water, mixing the lens of fresh water with the salt water below.

The human population of the Bahamas has a long history. The first visitors were probably the hunter-gathering Ciboney, meaning 'people of the precious stone', who drifted between Cuba and Florida in 500 BC, but lived

OPPOSITE An aerial view of some of the low-lying islands and sandbanks of the Bahamas showing the flow of the ocean currents.

on Cuba. It was the Lucayan Amerindians, Arawak from South America, who first settled here in about AD 500. They were harassed regularly by the warmongering Caribs from the south and in 1492 unwisely welcomed Columbus, prompting the European invasion and their own demise.

The environment was hit hard by the invasion too. As the Europeans came in, the trees went out. Valuable timber was carried back across the Atlantic. The Lucayans had used mahogany sparingly for dugout canoes, but post-Columbus the mature trees vanished almost overnight. Many years later, the British finished off the pine forests, the trunks used for masts by the Royal Navy.

Timber was not the only item imported to Europe. If one theory describing the disease's origins is to be believed, the Spanish explorers also introduced syphilis. The first recorded outbreak in Europe was in Naples in 1494, just after Columbus had returned from his initial voyage. Fortunately, they also brought back the so-called 'sailor's cure', a decoction from the bark of the gaiac or ironwood tree *Lignum vitae*, meaning 'tree of life'. It contains saponins, which have proved effective against the spirochete bacteria *Treponema pallidum* that cause syphilis; but that is not all. The tree's sap or resin has anti-inflammatory properties and is used to treat gout and arthritis, and tea made from its blue flowers and bright-green leaves is reputed to be an excellent energy restorative. The wood is also the densest and hardest known in the world – it sinks in water and resists rot, so for a variety of reasons it too was exploited and commanded high prices in Europe in the early 1500s. Bahamian sources were quickly exhausted.

The Lucayans, like their trees, declined rapidly in numbers. Those who were not killed or enslaved succumbed to European diseases; but some of their legends live on, like the mythical beast of Andros, the so-called 'chickcharnie' or 'yahoe'. It had a body like a bird, three toes on each foot, bright-red eyes, and stood about 1 metre (3 feet) high; it might well have been based on a real animal, one that has only recently been identified. Erosion of the island's limestone surface has resulted in holes filled with soil; known as 'banana holes' because they are good for growing bananas, they also contain all manner of palaeontological treasures. The remains of ancient iguanas and tortoises have been unearthed here, but more significant was the skeleton of a giant barn owl *Tyto pollens*; the other bones had belonged to its prey.

The chickcharnie owl was the principal predator of a creature that is still with us, but only just – the Bahamas hutia *Geocapromys ingrahami*, the only terrestrial mammal ever to reach these islands and one that is now on the brink of extinction. In fact, until 1966, it was thought to be extinct, but on the tiny island of East Plana Cay each evening up to 10,000 emerge from

Lignum vitae, meaning 'tree of life', has the densest and hardest wood known in the world – it sinks in water

their daytime refuges among the rocks to graze on low vegetation. It may seem a healthy number, but this is the entire world population of this particular species (although it is related to hutias on Cuba and Hispaniola) and should disease or some other disaster such as a hurricane strike, they could disappear in an instant. Today, their main enemies are domestic and feral dogs and cats, so to ensure a future for the species, some have been transplanted to other islands with some success. One site is Exuma Cays, location in 1958 of the Bahamas' first national park.

The hutia's natural surviving predators are boid or constrictor snakes *Epicrates* spp., of which several species are found throughout the islands. They substitute for mammal predators, taking roosting birds and bats, although they themselves fall foul of cats and dogs. Other endemic reptiles include the Bahamian rock iguana *Cyclura carinata* and the Cat Island terrapin *Trachemys terrapin*, and there are numerous local subspecies isolated on the islands with close relatives found elsewhere in the West Indies.

Take Bimini, for example, a small island in the northern part of the group. It has the Bimini boa *Epicrates striatus fosteri*, described in 1941 by Harvard's Thomas Barbour as 'black as a raven's wing – with a glittering pearly iridescence of extraordinary beauty'. It can reach 2.5 metres (8 feet) long if allowed to grow, but in the wild today few achieve this size. Many have been collected for the pet trade, while building developments in wilderness areas threaten the snake's very existence.

There is a Bimini Island ground boa *Tropidophis canus curtus*, which also occurs on New Providence and Cay Sal Bank and has the surprising trick of bleeding from the eyes when attacked. More common is the Bahamian brown racer *Alsophis vudii picticeps*, found throughout the archipelago, but Bimini has its own subspecies. It is a mildly venomous back-fanged snake with a penchant for anole lizards, but it poses little hazard to humans. Two blind snakes – the more common Bahamian pink blind snake *Typhlops biminensis*, which is often mistaken for an earthworm, and the less widespread Bahamian brown blind snake *T. lumbricalis*, which lives mainly on the larger islands – complete Bimini's snake tally. Lizards are represented by several species of the ubiquitous anole lizards, together with ameivas, geckos and curly-tails.

ABOVE The diminutive Abaco Island boa *Epicrates exsul* grows to 80 cm (30 in) long.

Birds, despite human disturbance past and present, play a leading role in the life of all the islands. Once upon a time, there were parrots everywhere.

Columbus noted in his log that 'flocks of parrots darken the sky'. The Lucayans kept them as pets and traded them with the Europeans, but when the land was cleared and new crops introduced, the birds first lost their living space and then their lives; they were persecuted for damaging the harvest. Today, just two small populations of the Bahama parrot *Amazona leucocephala bahamensis* survive, one on the island of Abaco and the other on Great Inagua.

The Inagua population nests in trees, like most parrots, but the Abaco birds are unusual in that they nest in subterranean hollows – limestone-solution cavities – the only large parrots in the New World to do so. It is a strategy that has proved useful on at least one occasion, although the advantage was short-lived. On 10 July 2004, fire swept through a nesting site and destroyed most of the 77 active nests, yet all the unfledged chicks and their brooding parents were unhurt – they were safe in their underground hollows. The surrounding Caribbean pine and

ABOVE One of Abaco Island's Bahamas parrots, a species confined to just two islands in the Bahamas group.

broadleaf coppice, however, was burned to the ground, making access easier for feral cats and introduced raccoons, and about 50 chicks were killed and eaten. And, if that were not enough, on 3 September – just as the chicks in the 56 surviving nests were about to fledge – Hurricane Frances hit the island with 320-km/h (200-mph) winds and lashing rain, followed just three weeks later by Hurricane Jeanne. The surviving population is on a knife edge.

Equally threatened is the white-crowned pigeon *Columba leucocephala*, a species that island-hops as far north as the Florida Keys in search of fruiting trees or nesting sites. These pigeons nest in the mangroves but are unproductive breeders, each pair raising only one chick each year – which is a problem, for thousands of them are shot annually by local hunters. The birds barely keep pace with the demand, although nowadays a hunter's daily bag in the Bahamas is restricted to 50 pigeons, and there is a hunting ban between 1 March and 28 September.

On Great Inagua, hyper-saline Lake Windsor is the site of an enormous colony of Caribbean flamingos *Phoenicopterus ruber ruber*, the national bird of the Bahamas. About 50,000 – almost half the world's entire population – arrive here each spring to breed and feed, then in winter they head for the inland salt lakes on Hispaniola. The Bahamians were once not as kind to their avian visitors as they are now. As recently as the 1950s, there was a permitted annual round-up of young flamingos that provided meat and decorative feathers.

ABOVE A flock of Caribbean flamingos gathers at the water's edge near a nesting site in the Bahamas.

These are just three of 304 species of birds recorded on the islands. Many are residents, but a fair number of migrants have adopted the islands either as a winter refuge or as a refuelling stop. The pectoral sandpiper *Caladris melanotos*, for example, stops off on its extraordinary annual migration between Argentina, where it spends the winter, and northern Canada, where it breeds in summer. Such is the importance of the wilderness areas of the Bahamas that if they were to disappear the sandpiper would fail to complete its journey.

The undoubted wildlife celebrities around here, though, are not birds but dolphins. The Little Bahama Bank is the playground of several species – notably the Atlantic spotted dolphins *Stenella frontalis* – and the dolphin tourist boats that seek them out. As their scientific name suggests, these dolphins have a long, narrow 'beak' and, as their common name indicates, they are covered with spots; at least the adults are – youngsters are plain grey at first and gradually acquire their 'spotty' appearance as they grow. They have their own language of whistles and burps, each dolphin with its own signature whistle. They use echolocation to find food, and the Bahamas spotted dolphins have learned how to zap the seabed and locate small fish hidden underneath; then they 'crater dive' with their snout into the sand to catch them, a behaviour that has become known as 'crater feeding'.

ABOVE A spotted dolphin examines the seabed with its echolocation system. If a fish is hiding under the sand, the dolphin will 'crater dive' to reach it.

The dolphins are just one of the underwater delights for visitors. As with all the Caribbean islands, many of the pristine wildlife sites in the Bahamas are in the sea, and they are easy to find. Not far from the capital, Nassau, on the island of New Providence, for example, the remarkable diversity of underwater life is there for anyone with a mask and snorkel to see. Close to the beaches are sea-grass beds, where the peacock flounder *Bothus lunata* and queen triggerfish *Balistes vetula* live. Rocky and coral patches, including 'bommies' – roughly circular plateaus of coral surrounded by sand – together with discarded tyres and oil drums, provide hiding places for the rock beauty *Holocanthus tricolor* and jackknife fish *Equetus lanceolatus*. Fringing coral reefs have shoals of French grunt *Haemulon flavolineatum*, which hide in the shade of an overhang and engage in 'kissing displays' at spawning time. Iridescent blue chromis *Chromis cyaneus* swim over the open reef, while hogfish *Bodianus* spp. hover close to an escape hole in case of danger.

The sand flats are the domain of the southern stingray *Dasyatis americana*, which feeds on crabs, worms, shrimps and shellfish. It finds its prey by ultra-sensitive receptors in its flattened snout and then covers it with its body, consuming its meal while buried in the sand. High on the list of preferred foods is the variety of large gastropod molluscs or sea snails with which it shares the reef. The helmet *Cassis flammea* is one of several species from which cameos are carved. Larger species, such as the king helmet *C. tuberose*, feed on sea urchins that they engulf and inactivate with a poisonous enzyme. They then dissolve the shell with an acid secretion to get at the soft organs inside. Tun shells of the family Tonnidae have a giant foot with which they envelop sea cucumbers, and the flamingo tongue shell *Cyphoma gibbosum* feeds exclusively on delicate sea fans. The giant queen conch *Strombus gigas*, by contrast, is a vegetarian and feeds on sea grass. Large banks of conch shells in Nassau harbour, however, are testament to its over-exploitation for its meat, the main ingredient in

many Caribbean dishes. Nevertheless, it does survive and can live for up to 25 years, if it does not fall prey to eagle rays *Aetobatus narinari,* which can crush it with a single bite, or the diminutive tulip shell *Fasciolaria* spp., which bores through its shell and eats the conch from the inside out.

Sea cucumbers have a hard time too. Many are attacked by the juvenile pearlfish *Encheliophis* spp., which gains entry via the cucumber's anus. The fish lives inside its unfortunate host and feeds on its gonads. As sea cucumbers continually replace their organs, the parasite is guaranteed a daily meal while growing up. Some pearlfish do not leave when they mature but spend almost their entire lives in the anus of sea cucumbers, not parasitizing their host but simply hiding inside by day and emerging at night to feed. As the sea cucumber breathes through its anus, the pearlfish is always guaranteed passage in or out. Other pearlfish find refuge inside the shells of conches and clams.

At the drop-off into deeper water (only about 100 metres/300 feet from the shore on the west coast of New Providence), damselfish *Stegastes* spp. patrol the rich growth of corals, and black-capped basslets *Gramma melacara* occupy every crevice. They emerge cautiously to forage along the reef edge, swimming upside down beneath overhanging rocks and corals. Over the deep blue, squadrons of manta rays *Manta birostris* glide effortlessly, looping the loop while filtering plankton from the surface waters. A lone and inquisitive 2-metre (6½-foot) long great barracuda *Sphyraena barracuda* might inspect newcomers, whoever and whatever they may be. He is top fish around here.

ABOVE A princess helmet Cassis flammea uses its enormous foot to engulf and then feed on a sea urchin.

OPPOSITE A squadron of spotted eagle rays swims in the vicinity of the Bahamas.

ABOVE A female lemon shark gives birth to her pups in the horseshoe-shaped lagoon in Bimini. The newborn pups head immediately for the mangroves.

This is also the realm of the wall diver. The Bahamas are well known for extensive and diverse underwater walls, like the famous Andros Wall at the edge of the Tongue of the Ocean (known locally as TOTO), the 1800-metre (5900-foot) deep abyss between the islands of Andros and New Providence. It is here that the legend of the lusca has its origins. She is – or at least is supposed to be – a mythical giant octopus (half octopus and half shark in some accounts) that pulls fishermen down into a watery grave. She lives in the blue holes, and her breathing accounts for the tidal flow of water in and out of the caves.

There are also sharks here. The great hammerhead *Sphyrna mokarran*, recognized by its tall dorsal and short pectoral fins, skims the sand searching for its favourite prey – stingrays. One individual was seen with a half a dozen barbs firmly implanted round its mouth. The tiger shark *Galeocerdo cuvier*, distinguished by the faint stripes on its side, eats just about anything, earning itself the nickname 'garbage collector of the sea'. Contrary to their ferocious reputation, tiger sharks are timid, taking time to come into baits left by divers. Schools of blacktip

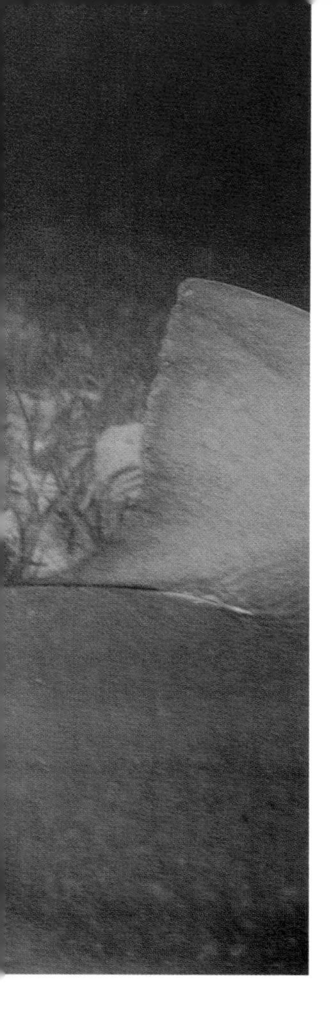

sharks *Carcharhinus limbatus* patrol the drop-off, taking full advantage of any fish unwise enough to leave the safety of the reef, while the stocky bull shark *C. leucas* frequents artificial reefs and wrecks. At one dive site, Shark Canyon, there are even 'friendly' nurse sharks *Ginglymostoma cirratum* that will present themselves to divers in the hope of getting a scratch.

The most studied shark in the Bahamas must be the lemon shark *Negaprion brevirostris*. At Bimini, researchers from the University of Miami have recorded just about every aspect of their life. They have seen how the female sharks enter the horseshoe-shaped lagoon and give birth to their pups. These then make straight for the mangroves, where they spend the first year or so of their life feeding on small fish, crustaceans and octopuses. When they first move into open water, they head for parts of the lagoon where others of their own size congregate. As they grow, they become gradually more adventurous, eventually leaving the safety of the lagoon altogether. Elsewhere in the Bahamas, especially during the winter months, row upon row of them rest closely together on the sand, one piece of behaviour that is yet to be explained.

One of the most common sharks here is the sleek and agile Caribbean reef shark *Carcharhinus perezi*, seen mostly close to healthy reefs, where its main prey is fish. It is the shark most likely to turn up during a shark-feeding circus, a controversial practice but undoubtedly an exciting one. Shark attacks on humans are rare but not unknown, and shark feeding is thought to encourage sharks to associate us with a free meal. With more people in the water these days, it is inevitable that the sharks are going to make a mistake or two, and those, of course, become headline news. The reality, according to the International Shark Attack File, is that there have been no more than 25 attacks in Bahamian waters between 1749 and 2005, of which one was fatal. In the Antilles during the same period there have been 38 attacks and 18 fatalities, hotspots being Cuba and Puerto Rico. So the chance of a shark attack off the Bahamas is way down the list of dangers. Falling coconuts are a greater risk.

Anyway, sharks feed mainly at night. At sunset, the reef comes alive as the night shift takes over from the day-trippers; and while the parrotfish are tucked away in the mucus sleeping bags that prevent their smell from giving away their position to sharks, the squirrelfish and black-bar soldierfish emerge from their daytime nooks and crannies to feed, and lobsters and crabs break cover to forage

ABOVE Caribbean reef sharks at Walker's Cay in the Bahamas.

RIGHT The tiger shark, recognised by the stripes on its sides, favours murky coastal waters, and is frequently found in the shallow waters close to island chains, including lagoons.

The chance of a shark attack off the Bahamas is way down the list of dangers. Falling coconuts are a greater risk

over the open sand flats. The nocturnal fish are agile and quick to take flight, wary of reef sharks and moray eels scouring the coral heads, but as they swim and disturb the water, the darkness is sometimes interrupted by the bioluminescent flashes of 'sea sparkle', product of the microscopic dinoflagellate *Noctiluca scintillans*.

This tiny alga has a couple of flagella, but it does not use them for locomotion. The only movement it can achieve is up and down, the result of buoyancy adjustments caused by changing the ionic content of its cytoplasm. This complicated science may seem an irrelevant aside, but it is sobering to realize that it has great significance to the health of the reef. Early in the season *Noctiluca* is negatively buoyant and distributed throughout the water column, but later the cell accumulates ammonia and becomes positively buoyant. The bloom then collects at the surface to form a 'red tide', and this can kill certain marine organisms, such as fish, or accumulate in the tissues of others, such as mussels and oysters. The end result for us is paralytic shellfish poisoning and a trip to the hospital.

If the time is an hour after sunset on the second day after the full moon, dinoflagellates could well be joined by the Bahamian fireworm *Odontosyllis* spp., which produces light, not only as a startle response like the dinoflagellates but also when spawning. The worms normally live in tubes on the sea floor, but at spawning time they float to the surface, where the females flash a phosphorescent green glow that attracts the males; the males then flash too until each pair releases its eggs and sperm, flashing as they go. One theory proposes that the 'mysterious lights' in the ocean seen by Columbus on the night before he landed in the Americas were from these creatures. That night – 10 October 1492 – would have been in the third quarter of the moon, one of the times when fireworms spawn.

Columbus, it is claimed by the neighbouring Turks and Caicos Islands, landed on Grand Turk in 1492; at least, a monument and plaque in a small plaza surrounded by the government buildings in Cockburn Town say so. Grand Turk is treeless and brush-covered, and one of eight inhabited islands in an archipelago that is, in point of fact, a southeast extension of the Bahamas reef and bank system, an undersea formation that continues to Silver Bank and Christmas Bank off the coast of the Dominican Republic (see chapter 2).

The name 'Turk' was once thought to refer to the Turk's head cactus *Melocactus intortus* var. *intortus*, which can be found on the islands to this day, but research at the national museum suggests it had a more ominous origin. Pirates in the Mediterranean were once known as Turks, as Ottoman crews were responsible for piracy in that region. On French maps of the late seventeenth century, the Turk islands were labelled *Conciua ou Turks*, meaning 'where the Turks gather'; a warning to mariners that this was a danger area.

Many of the cays in this archipelago have pirate connections. French Cay on the Caicos Bank was headquarters for Nau L'Ollonais, one of the cruellest pirates to sail the Spanish Main. It was said that in a raid in Panama he sliced open the chest of a woman, took out her heart, ate it while it was still beating and washed it down with her blood. He was so despised that after his capture he was torn limb from limb and the pieces spit-roast on a bonfire. Parrot Cay, formerly known as 'Pirate Cay', was temporary home to Ann Bonny, who took refuge here having fled Jamaica after her lover Calico Jack Rackham was hanged.

Dominating Grand Turk are several *salinas* or salt ponds, source of the island's most important product until the industry collapsed in the 1960s. Nearby Salt Cay was the largest producer of salt in the world. Over a hundred vessels a year transported the 'white gold' to the USA and beyond. The ponds were linked to the sea by canals and sluice gates, water flow controlled by windmills. The windmills and salt sheds are still standing, albeit only just, a dilapidated window on nineteenth-century island life. The island was brought abruptly into the twentieth century in 1962, when astronaut John Glenn splashed down in Grand Turk waters after having become the first American to orbit the Earth.

The islands' very distant past is also coming to light in an archaeological excavation at Coralie, near the northwest tip of North Creek. It reveals the skeletal remains of now extinct 60-cm (2-foot) long native land tortoises that must once have gorged on cactus fruits, and land iguanas *Cyclura carinata* that are 30 cm (12 inches) longer than their modern descendants and, judging by the number of bones that have been discovered, must have had a density of many hundreds to the hectare (several hundred to the acre). There were also parrots galore, an indication that the island was probably once densely forested. Other finds include the bones of birds long gone from these shores – the scarlet ibis *Eudocimus ruber*, now found only in South America and Trinidad and Tobago, and the thick-knee *Burhinus* spp., a ground-dwelling bird whose only current home in the Caribbean is on Hispaniola. The flora and fauna of these islands must have been very different when these creatures were alive.

Providenciales is the most densely populated and tourist-developed island in the Turks and Caicos, although its western half is mainly wilderness, with high ridges covered with scrub and cactus. The natural monuments here are Chalk Sound National Park, with its turquoise water and numerous mushroom-like islets, and the Northwest Point Marine National Park that encompasses coral reefs and saline lakes, the latter a magnet for breeding and migrating waterfowl. It also has the only conch farm in the world.

Middle Caicos is the largest island in the group and is riddled with caves, including the Conch Bar Caves, the largest above sea level in the entire Bahamian islands region. It is the site of an unusual investigation by scientists who are studying stalactite formation. They are counting – wait for it – drips.

There were also parrots galore, an indication that the island was probably once densely forested

OPPOSITE An aerial view of the resort at Little Palm Island in Lower Florida Keys.

North Caicos has a huge osprey *Pandion haliaetus* nest at Third Cay, one of the famous Three Mary Cays; Caribbean flamingos feed on the tidal flat to the south of the island; and whistling ducks *Dendrocygna arborea* are seen at Pelican Point. It also has its own blue hole – Cottage Pond, surrounded by ferns and other tropical plants. It is said that, in days gone by, slaves were thrown into it and drowned, and today the ferns are in such a good condition that the locals say they are tended by spirits. The freshwater pond is an ancient sinkhole that 20 metres (65 feet) down opens into a salt water chamber. The upper layer of the pond has water stained red with tannins from the surrounding vegetation, and the lower level is clear salt water. The boundary is a thick, almost soup-like layer that has the 'bad eggs' smell of hydrogen sulphide. At 23 metres (75 feet), small tunnels branch off, leading to the sea, and there are stalagmites and stalactites. As ponds go, it is pretty impressive – the bottom of the main shaft is 70 metres (225 feet) below the surface.

The easternmost island in the group is South Caicos, much of it protected by the Admiral Cockburn Land and Sea National Park. The high point here, quite literally, is the Sail Rock Hills, 54 metres (178 feet) above sea level, from which there are views of Turks Passage, where divers are confronted by a 2000-metre (7000-foot) vertical wall, and Belle Sound, a vast bay that opens on to the Caicos Bank and is a fishing centre for bonefish.

Fishing – big-game fishing in particular – is the most popular activity in our last outer-Caribbean location, the Florida Keys. The Keys were home to writer Ernest Hemingway, a devotee of the 'sport' and an enthusiastic hunter. Hunting and fishing in the Hemingway style still continue as a national pastime, but in sanctuaries and parks they are prohibited.

Today, wildlife here is protected in the Florida Keys Wildlife and Environment Area, a long, thin, string of islands that stretches about 130 km (80 miles) from Key West to Key Largo. Each island is made of limestone, and tropical Caribbean and more temperate North American plant communities overlap. Significant habitats are the tropical hardwood 'hammocks' (thickly wooded land with bushes and vines), mangrove wetlands and coastal salt marsh. The hammocks are especially important, for many have been lost to urban development throughout Florida because they grow on the higher, drier land that is suitable for building. The survivors contain a who's who of tropical and semitropical trees and shrubs, over a hundred species having been identified. Considered to be one of the most endangered habitats on Earth, they are critical as stopover and feeding sites for bird migrants between the eastern half of North America and Central and South America. Among the residents, there are over 30 threatened animal species, of which several are found nowhere else in the world. The Key Largo cotton mouse *Peromyscus gossypinus allapaticola*, for example, has been a casualty of hammock destruction, while the silver rice rat *Oryzomys palustris natator* has lost much of its wetland habitat. Both are endangered.

The key or toy deer *Odocoileus virginianus clavium* is the smallest subspecies of

the white-tailed deer and present on only 26 islands from Big Pine Key to Sugarloaf Key. It is estimated that it munches on over 160 species of plants, including mangroves, and its greatest threat these days is not from hunters but from automobiles: 70 per cent of deaths are road kills.

The Lower Keys marsh rabbit *Sylvilagus palustris hefneri* has an unexpected sponsor – Hugh Hefner of Playboy Corporation fame. The billionaire supported research on the species when it was declared endangered by the US Fish and Wildlife Service, and he was honoured by having the subspecies named after him. Subsequently, this particular marsh rabbit has been known affectionately as the 'playboy bunny'. Domestic and feral cats are its main problem, coupled with the inevitable habitat destruction and road kills.

Of the snakes, the big pine or keys ring-necked snake *Diadophis punctatus acricus* lives in hardwood hammocks and among limestone outcrops, and is on Florida's threatened-animals list as it is found on one island only. It has the unpleasant habit of secreting musky, pungent-smelling saliva from the corners of the mouth, but it still falls prey to other snakes, feral hogs, owls, opossums, skunks and even bullfrogs. The non-venomous southern black racer *Coluber constrictor priapus* is more successful. It is, as its name suggests, black and fast, and it has a catholic diet, its large eyes with chocolate-brown irises betraying a hunter that uses vision. It is active during the day in scrub near water and often raises its body cobra-fashion to peer over the low vegetation.

Some of the 'lowlier' residents are also worth a mention. The beautiful Stock Island tree snail *Orthalicus reses reses* is found only in the tropical-hardwood hammocks in the extreme south of Florida and the Keys. More closely related to snails in the West Indies, it is likely that this species rafted here from Cuba or Hispaniola and that each isolated population then developed into its own colour form; more than 50 exist today, although there were many more that have become extinct, the result of over-zealous collecting.

Of the myriad insects, the fabulous Schaus swallowtail butterfly *Heraclides aristodemus ponceanus* stands out from the crowd. It was once found all over Florida, but the disappearance of its hardwood-hammock habitat has turned it into one of the rarest butterflies in the area. Spraying with insecticides and butterfly collecting have had an impact, but the worst blow was Hurricane Andrew in 1992, which left just 70 individuals in its wake. Fortunately, eggs had been collected and a captive-breeding programme instigated. Linked with habitat restoration, especially of its torchwood *Amyris elemifera* food plant, the butterfly is making a comeback, albeit slowly.

The big pine snake has the unpleasant habit of secreting pungent-smelling saliva from the corners of the mouth

Of the birds, large breeding colonies of white ibis *Eudocimus albus* are conspicuous from the air, but isolated from the human populations in remote wetland areas. They nest from March through August, and then range widely to sites around the southeastern USA and Greater Antilles. Seen in large flocks, flying in long lines or V-formations, they certainly deserve the accolade of being the bird symbol of Florida. When feeding, ibises stab into the mud, picking out small aquatic life such as crayfish, worms, insect larvae, frogs and fish. The mainly pink roseate spoonbill *Ajaia ajaja*, on the other hand, uses the sensitive nerve endings inside the tip of its long, spoon-like bill to sweep through the water and mud, and then snap up any organisms it touches. The bird was once hunted for its feathers. At some nesting sites, the roseate tern *Sterna dougallii* has shunned isolation in favour of gravelled roof tops on condominiums and dredged-material islands, whereas colonies of the little blue heron *Egretta caerulea* (a close relative of the ubiquitous snowy egret *E. thula*) prefer isolated islands to build their platforms of sticks in trees and shrubs.

The seashore and its inhabitants are protected by the Florida Keys National Marine Sanctuary, the most extensive marine reserve in the USA and North America's only barrier reef – the third largest in the world after the Great and Belize barrier reefs. Some of the corals, such as the boulder types, grow at less than 1 cm (½inch) per year, while the branching corals extend by 10 cm (4 inches). As on other barrier reefs the coral spawn synchronously, in this case five to eight days after the August full moon. Soft corals, such as sea whips and sea fans, predominate, and it is claimed that almost every major group of animals on our planet is represented here.

A second important marine community is supported by the sea-grass meadows. Sea grasses are flowering plants that, by day at least, produce oxygen, and in Florida they cover an estimated 1 million ha (2.7 million acres), helping to keep the water clear, stabilizing bottom sediments and providing a living space for marine organisms, food for animals, such as conches and manatees, and a nursery area for young fish and invertebrates. Similarly, Florida's 200,000 ha (496,000 acres) of mangroves form protective nursery areas for fish, crustaceans and shellfish, and provide food for economically important fishes such as tarpon, snook, snapper and jack, as well as oysters and shrimp.

Among the monsters in the sea-grass beds and on patch reefs is the Florida horse conch *Pleuroploca gigantea*, one of the largest spindle-shelled gastropod molluscs in the sea and Florida's designated state shell. It can grow to 60 cm (24 inches) long and feeds on other gastropods, such as tulip shells *Fasciolaria* spp., lightning whelks *Busycon* spp. and lace murex *Chicoreus* spp., grasping the victim's operculum (the flap covering the opening of the shell) to prevent it closing up, and inserting its proboscis so that it eats away at the victim's soft tissues inside.

At the southernmost tip of the Keys is the Key West National Wildlife Refuge, established in 1908 by President Theodore Roosevelt. It is a small reserve, but

important to 250 species of birds, including the piping plover *Charadrius melodus* and a white morph of the great blue heron *Ardea herodias*, known locally as the great white heron, and three species of sea turtles. Green *Chelonia mydas*, loggerhead *Caretta caretta* and hawksbill *Eretmochelys imbricata* turtles nest here, albeit in relatively small numbers on the narrow beaches.

About 100 km (70 miles) west of Key West is the Dry Tortugas National Park, seven islands consisting of coral reefs and shifting sands, plus an enormous fort – Fort Jefferson – one of the largest coastal forts in the world, built with over 6 million hand-made bricks. The islands were given their name in the early sixteenth century by Ponce de León. Mariners stopping here at that time provisioned their ships with the meat of the abundant sea turtles – *tortugas* in Spanish. There was, however, no water, so the islands became known as the Dry Tortugas.

They are at the confluence of the Caribbean Sea, Gulf of Mexico and Atlantic Ocean, and so have been a prime site for shipwrecks. In 1742, the British warship HMS *Tyger* went aground on Garden Key, marooning 242 sailors. They survived for two months by eating turtles, monk seals and fish, and finally returned to Jamaica in a sloop they built from their ship's wreckage, hijacking a Spanish schooner on the way. Another 200 wrecks are scattered around the national park, including the remains of the windjammer *Avanti*, which went down in January 1907 off Loggerhead Key while carrying timber from Pensacola to Montevideo. Today, it is a refuge for an extraordinary number of reef fish: during surveys in the mid-1970s up to 135 species were recorded.

The reefs themselves are as pristine as they were decades ago. Lobsters, crabs, clams and conch spawn here, and the ocean currents sweep their larvae northwards to replenish ecosystems along the Keys and the southeast coast of the USA.

On land, these islands become crowded in spring and autumn when passage migrants – warblers, vireos and numerous other songbirds – from South and Central America put down to rest. West Indian species such as loggerhead kingbirds *Tyrannus caudifasciatus* and Bahama swallows *Tachycineta cyaneoviridis* turn up from time to time, and on the smaller and less inaccessible keys, such as Long Key, magnificent frigatebirds *Fregata magnificens*, masked boobies *Sula dactylatra* and brown noddies *Anous stolidus* have their nests. Up to 100,000 sooty terns *Sterna fuscata* are found on Bush Key, the only regular breeding colony in the USA.

ABOVE Common or brown noddies arrive in early spring at Bush Key in the Tortugas, the only breeding site for this species in the USA.

OPPOSITE Populations of roseate spoonbills took a serious knock when the wings were turned into lady's fans during the early 1800s.

THE SEARCH CONTINUES

THE CARIBBEAN is a place that has seen much change in a short time. It had a beautiful people – the Arawak – who gave us such everyday terms as avocado, barbecue, canoe, hammock, potato and tobacco. It has given us a whole new cuisine of exotic food and drinks, music and dance – reggae and salsa – and the infectious gaiety of carnivals. It had pirates and buccaneers, and even today some still operate in no-go areas for tourist yachts. Some wilderness areas are under threat, where wildlife is barely hanging on, but there are still paradise islands and coral reefs that sparkle like jewels in an azure sea, enticing sun worshippers, divers and nature lovers alike; and over 15 million visitors to the Caribbean each year cannot be wrong!

BELOW Islands, palm trees, coral and white sand beaches – the quintessential Caribbean paradise.

In such a condensed account, however, many islands and cays have had to be left out, and some fall outside the area we have explored. Bermuda, that small dot way out in mid-Atlantic – a volcanic island surrounded by deep ocean – is too far away to count as 'Caribbean', although its influence on the region touches on commerce and even myth. It was Bermudan traders who dominated the salt industry in the Turks and Caicos and Bahamas, for instance, and the island sits at the apex of the infamous Bermuda Triangle, with the Bahama Banks at its base. Here ships and planes have mysteriously disappeared, and science is at a loss to explain why.

There are also baffling natural phenomena, like the huge 'bubble', about 2 km (over 1 mile) across, seen by the crew of a commercial airliner on the surface of the

Atlantic just a few minutes out of Puerto Rico in April 1963. Maybe the mind-boggling concept of 'oceanic flatulence', the release of methane gas from the seabed, explains the event; after all, it is one of the forces put forward to explain vanishing tricks in the Bermuda Triangle.

Then there are the equally puzzling 'whitings', 3-km (2-mile) wide patches of milky-coloured water resembling sandbanks that occur off the Bahamas. They are often seen in the summer months to the west of Andros and Abaco and on the Little Bahama Bank, and have appeared on Admiralty charts for hundreds of years. Today, they can be spotted on satellite and shuttle pictures. At first, it was thought that they were caused by shoals of fish, called 'whitin', but now we know they are clouds of calcium-carbonate mud suspended in water, possibly produced by blooms of cyanobacteria and unicellular green algae (although the jury is still out). There are larger fish present, however, and these are sharks – snow-white blacktip sharks. Why they live exclusively in these white clouds, and why they are coloured in this way has yet to be explained. One suggestion is that they stir up the bottom sediments to produce the white clouds, creating a kind of fish trap for sneak attacks in the same way that spiders spin webs to catch insects.

And, while we're on mysteries, there is a last word on Christopher Columbus, who wrote in his will that he would want to be buried in the West Indies. After he died at Valladolid, to the northwest of Madrid in Spain, on 20 May 1506, his remains were retained at the crypt of the convent of Observance in Valladolid until a suitable religious building could be found in the Caribbean. In 1513, they were moved to Cartuja de Santa Maria de la Cuevas, a cathedral in Seville, and in 1537 they finally crossed the Atlantic, to Santo Domingo, in what is now the Dominican Republic. In 1795, they were moved, for political reasons, to Cuba, and finally in 1899 back to Spain; at least, it is thought that they were moved. But in 1877, workers digging in Santo Domingo's cathedral uncovered a lead box with an inscription that translates as 'Renown man: Don Cristobal Colón'. The authorities deduced that the wrong man had been dug up in 1795, and that the genuine Columbus was still on the island. The upshot of all this is that both the Dominican Republic and Spain lay claim to his last resting place.

Recently, slivers of the bones in Spain were DNA tested and compared to the remains of Columbus's brother Diego. The results were positive, indicating that the memorial in Seville really does contain Christopher Columbus. His remains, however, were not complete. Now the pressure is on the Dominican Republic to open up their casket, which is entombed in a monument at Columbus lighthouse, and carry out similar tests. It could be, with all the confusion and constant moving, that part of the First Admiral of the Ocean in the Americas is in Europe and part in the West Indies!

If you are interested in finding out more and experiencing the wonders of the Caribbean for yourself, the gazetteer section that follows will be a useful start.

U.S.A
Fort Fauderdale
Miami
Bimini Islands
Key West
Florida Keys
Straits of Florida
Northwest Providence Channel
Northeast Providence Channel
Berry Islands
Great Abaco
Eleuthera Island
NASSAU
New Providence
Dive through a Hollywood set in New Providence's southwest corner, which offers a dozen deliberately sunk wrecks in close proximity
2
BAHAMAS
Andros Island
Santaren Channel
Cat Island
San Salvador
Conception Island
Rum Cay
Great Guana Cay
Long Island
Sandy Cay
Jumentos Cay
Crooked Island Passage
Samana Cay
Plana Cays
Mayaguana
Crooked Island
Acklins Island
Caicos Passage
TURKS & CAICOS ISLANDS (to UK)
Caicos Islands
Grand Turk Island
Turks Islands
Little Inagua
Great Inagua
ATLANTIC OCEAN
Climb over 150 routes in the lovely mogotes limestone scenery of the Viñales valley
1
LA HABANA (HAVANA)
Cay Sal
Canal Nicholás
Guanabacoa
Matanzas
Cárdenas
Sagua la Grande
Pinar del Río
San Antonio de los Baños
Santa Clara
Caibarién
Cienfuegos
Placetas
Morón
Cayo Romano
Canal Viejo de Bahama
Duncan Town
Isla de la Juventud
Sancti Spíritus
CUBA
Archipiélago de los Canarreos
Ciego de Ávila
Florida
Camagüey
Las Tunas
Holguín
Banes
Archipiélago de Jardines de la Reina
Manzanillo
Golfo de Guacanayabo
Sierra Maestra
Guantánamo
Windward Passage
Greater Antilles
CAYMAN ISLANDS (to UK)
Grand Cayman
Little Cayman
Cayman Brac
7680m
3
Pico Real del Turquino 1974m
Santiago de Cuba
GUANTANAMO BAY (to US)
Hispaniola
Île de la Tortue
Santiago de los Cabelleros
Cap-Haïtien
Gonaïves
Pico Duarte 3175m
San Francisco de Macorís
Puerto Rico Trench
PUERTO RICO (to US)
HAITI
DOMINICAN REP.
Canal de la Mona
Bayamón 1338m
SAN JUAN
Caguas
PORT-AU-PRINCE
La Romana
Mayagüez
Ponce
4
Pic la Salle 2680m
Barahona
SANTO DOMINGO
Isla Saona
Isla Mona
Isla Beata
Visit the Cayman Turtle Farm to see turtles at all stages of development from 10-cm (4-inch) long hatchlings to the grand 1-metre (3-foot) long adults
Montego Bay
Blue Mountain Peak 2256m
NAVASSA ISLAND (to US)
JAMAICA
Spanish Town
KINGSTON
Snorkel at night in Phosphorescent Bay through the natural light show that occurs when plankton are disturbed
Caribbean Sea
0 100 200 kilometres
0 100 200 miles

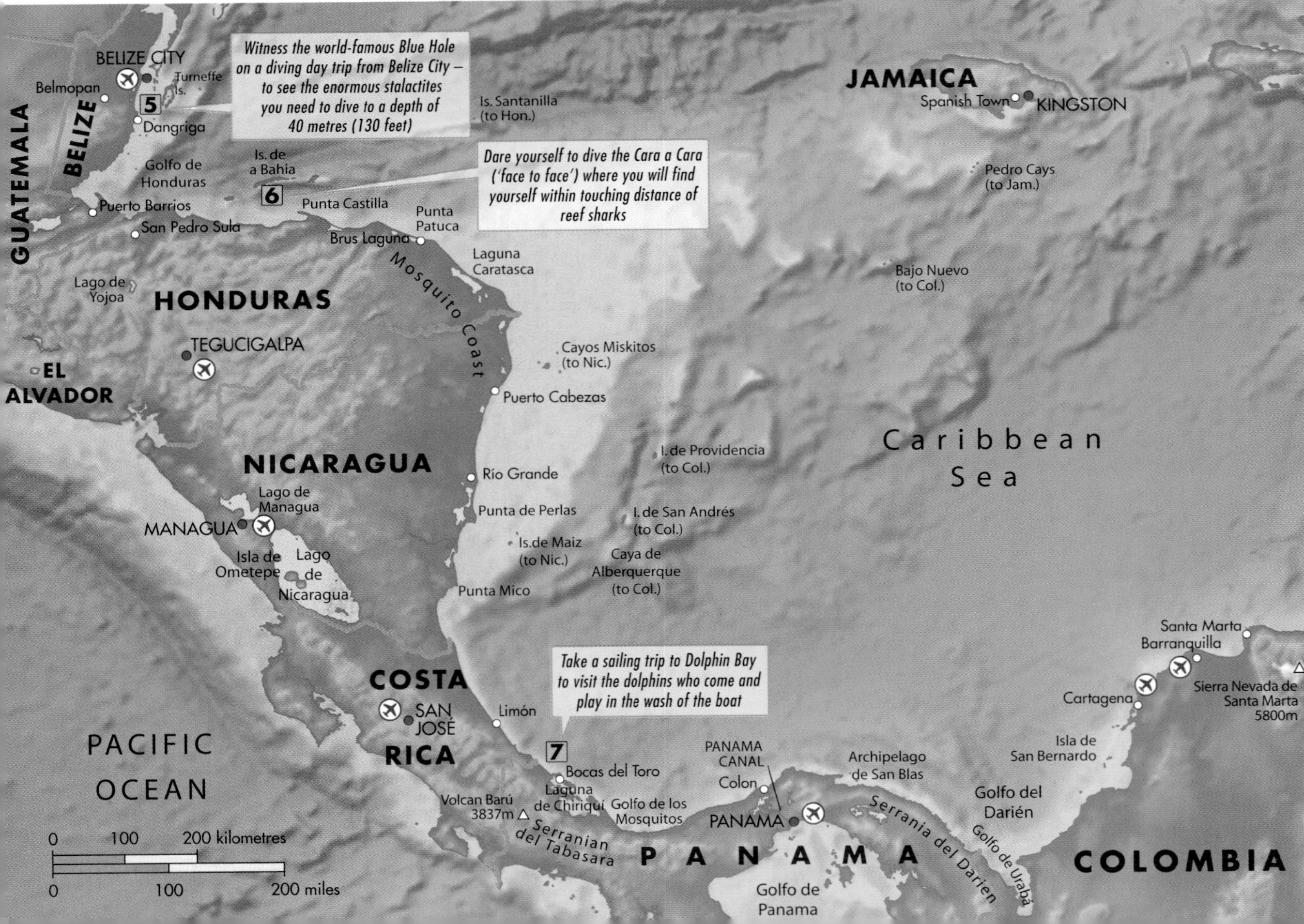

BELIZE CITY
Belmopan
Turneffe Is.
5
Dangriga
BELIZE
GUATEMALA
Witness the world-famous Blue Hole on a diving day trip from Belize City – to see the enormous stalactites you need to dive to a depth of 40 metres (130 feet)
Is. Santanilla (to Hon.)
JAMAICA
Spanish Town
KINGSTON
Golfo de Honduras
Is. de a Bahia
6
Dare yourself to dive the Cara a Cara ('face to face') where you will find yourself within touching distance of reef sharks
Pedro Cays (to Jam.)
Puerto Barrios
Punta Castilla
Punta Patuca
San Pedro Sula
Brus Laguna
Laguna Caratasca
Mosquito Coast
Lago de Yojoa
HONDURAS
Bajo Nuevo (to Col.)
TEGUCIGALPA
EL SALVADOR
Cayos Miskitos (to Nic.)
Puerto Cabezas
NICARAGUA
I. de Providencia (to Col.)
Caribbean Sea
Río Grande
Lago de Managua
MANAGUA
Punta de Perlas
I. de San Andrés (to Col.)
Is.de Maiz (to Nic.)
Isla de Ometepe
Lago de Nicaragua
Caya de Alberquerque (to Col.)
Punta Mico
Santa Marta
Barranquilla
Take a sailing trip to Dolphin Bay to visit the dolphins who come and play in the wash of the boat
COSTA RICA
SAN JOSÉ
Limón
Cartagena
Sierra Nevada de Santa Marta 5800m
PACIFIC OCEAN
7
Bocas del Toro
PANAMA CANAL
Colon
Archipelago de San Blas
Isla de San Bernardo
Golfo del Darién
Laguna de Chiriquí
Golfo de los Mosquitos
Volcan Barú 3837m
Serranian del Tabasara
PANAMA
Serrania del Darien
Golfo de Urabá
COLOMBIA
Golfo de Panama
0 100 200 kilometres
0 100 200 miles

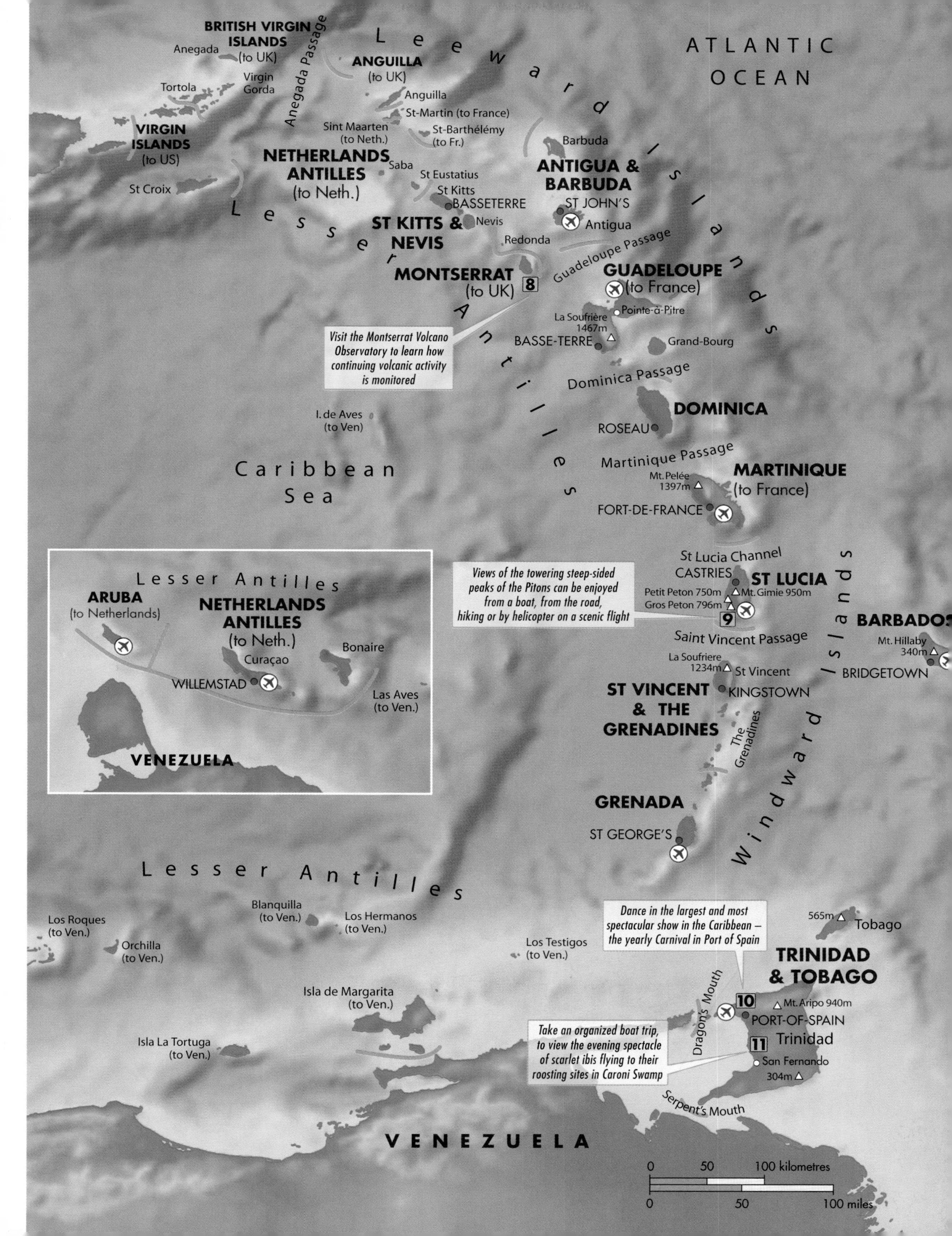

ATLANTIC OCEAN
Leeward Islands
Lesser Antilles
Caribbean Sea
Windward Islands
BRITISH VIRGIN ISLANDS (to UK)
Anegada
Tortola
Virgin Gorda
Anegada Passage
VIRGIN ISLANDS (to US)
St Croix
ANGUILLA (to UK)
Anguilla
St-Martin (to France)
Sint Maarten (to Neth.)
St-Barthélémy (to Fr.)
NETHERLANDS ANTILLES (to Neth.)
Saba
St Eustatius
St Kitts
BASSETERRE
ST KITTS & NEVIS
Nevis
Redonda
Barbuda
ANTIGUA & BARBUDA
ST JOHN'S
Antigua
MONTSERRAT (to UK)
8
Guadeloupe Passage
GUADELOUPE (to France)
Pointe-à-Pitre
La Soufrière 1467m
BASSE-TERRE
Grand-Bourg
Visit the Montserrat Volcano Observatory to learn how continuing volcanic activity is monitored
Dominica Passage
DOMINICA
ROSEAU
I. de Aves (to Ven)
Martinique Passage
MARTINIQUE (to France)
Mt. Pelée 1397m
FORT-DE-FRANCE
St Lucia Channel
CASTRIES
ST LUCIA
Petit Peton 750m
Mt. Gimie 950m
Gros Peton 796m
9
Views of the towering steep-sided peaks of the Pitons can be enjoyed from a boat, from the road, hiking or by helicopter on a scenic flight
Saint Vincent Passage
BARBADOS
Mt. Hillaby 340m
BRIDGETOWN
La Soufriere 1234m
St Vincent
KINGSTOWN
ST VINCENT & THE GRENADINES
The Grenadines
GRENADA
ST GEORGE'S
Lesser Antilles
ARUBA (to Netherlands)
NETHERLANDS ANTILLES (to Neth.)
Bonaire
Curaçao
WILLEMSTAD
Las Aves (to Ven.)
VENEZUELA
Blanquilla (to Ven.)
Los Hermanos (to Ven.)
Los Roques (to Ven.)
Orchilla (to Ven.)
Los Testigos (to Ven.)
Dance in the largest and most spectacular show in the Caribbean – the yearly Carnival in Port of Spain
565m
Tobago
TRINIDAD & TOBAGO
Isla de Margarita (to Ven.)
Dragon's Mouth
10
Mt. Aripo 940m
PORT-OF-SPAIN
Trinidad
11
Isla La Tortuga (to Ven.)
Take an organized boat trip, to view the evening spectacle of scarlet ibis flying to their roosting sites in Caroni Swamp
San Fernando
304m
Serpent's Mouth
VENEZUELA
0 50 100 kilometres
0 50 100 miles

GAZETTEER

THE CARIBBEAN is a region of such incredible variety, there is something for everyone. Apart from being the ultimate destination to chill out, it also offers ample opportunities for those who like to be active on holiday. Beyond whale-watching and birding, there's diving, caving, climbing and wonderful walking trails.

This gazetteer is not meant as a comprehensive guide to the whole area but rather a personal selection of top spots for those interested in the wilder side of the Caribbean. Compiled by the producers of the series, it represents some of the best places to see the wildlife and natural landscapes of this diverse region. Some of the islands mentioned in the book have deliberately been omitted, either because they are too 'touristy' for the true nature lover, or because they are politically or geologically unstable. But there should be plenty here to satisfy most people. Contact addresses and useful websites are given at the end of each section if you want to find out more.

CUBA

The jewel in the Caribbean's crown as far as wildlife conservation is concerned, Cuba has a higher percentage of protected areas than any other nation on earth. With around 360 species of birds, it is a mecca for bird-watchers, but it boasts hugely diverse habitats and scenery – and great diving sites too. Cuba is well served by international flights, but – unlike most other Caribbean countries – does require a visa for visitors from most western countries. Internal flights need to be booked in advance, though rail travel and car hire are both adequate, and independent travel is easy. Most resorts are well developed and have a good selection of accommodation options. Permits are required for some national parks, usually for a nominal fee that goes to park maintenance; most can be bought on the spot. Local guides are plentiful and can be arranged through such schemes as www.birdingpal.org.

ACTIVITIES

Bird-watching

- Ciénaga de Zapata Biosphere Reserve, only a couple of hours from Havana, is a vast wetland with a mix of mangroves, marsh and patches of forest and grassland. Its most famous resident is the bee hummingbird or zunzuncito, the world's smallest bird.
- The national park at Peninsula de Guanahacabibes at the far western tip of Cuba has good birding trails through coastal forest. The trail to La Cueva de las Perlas is good for Cuban trogons and Cuban parrots.
- The province of Pinar del Rio in western Cuba is also good, especially in La Guira National Park, Las Terrazas Biosphere Reserve and Soroa.
- Other areas include Sierra de Najasa, Gran Piedra and Alejandro de Humboldt Biosphere Reserve.

Walking and hiking

Cuba's fantastic variety of scenery can best be appreciated from the many walking and hiking trails around the island. The majority require a guide.

- Some of the most accessible nature trails are around Las Terrazas, about 50 km (30 miles) west of Havana in the Sierra del Rosario.
- From Mil Cumbres, in the west of Sierra del Rosario, it's possible to climb El Pan de Guajaibón, which, at 692 metres (2270 feet), is the highest point in western Cuba.
- In the far west of Peninsula de Guanahacabibes, the Cuevas de las Perlas trail is a gentle 45-minute walk to some collapsed caves, which are lit by sunlight and have trees and vegetation growing amongst the rock. These caves are also home to hutias, endemic rodents of Cuba, although they are hard to spot.
- Day tours of hiking and swimming are available among the mountains and waterfalls of Topes de Collantes, Valle de los Ingenios or Cascada El Cubano. It's also possible to hire horses around this area.
- In eastern Cuba the Gran Piedra Biosphere Reserve and Pinares de Mayari have a number of marked trails through forests, offering views of waterfalls and mountains.
- One of the best hikes on the island takes you to the top of Cuba's highest peak, Pico Turquino, in Gran Parque Nacional Sierra Maestra. The hike takes a minimum of six hours on the way up and at least four coming down, depending on the condition of the trail. You may prefer to overnight at a shelter – mountain refuges can be organized through such companies as Responsible Travel – or make a three-day hike of it from Las Cuevas to Alto de Naranjo. Sierra Maestra is a great area for waterfalls, local endemic flora, birds and reptiles.
- Guantânamo in the far east of the country has trails to El Yunque near Baracoa Bay and through the Alejandro de Humboldt National Park.

Rock climbing

- The Viñales valley, near the western tip of Cuba, is a well-known destination for climbers, with over 150 routes set in lovely mogotes limestone scenery.
- Other areas for climbing include Los Paredones, Valle de Yumuri and Escaleras de Jaruco on the borders of the Havana province, all of limestone.

Canyoning

- Topes de Collantes on the south coast of the island offers dramatic scenery, spectacular waterfalls and easy access – a good place for experienced canyoners and novices alike.

Caves and caving

- Cuba has more than 20,000 caves with varying degrees of accessibility, from Cueva San Martín Infierno, which has the world's largest stalagmite, to Cueva de los Peces near Playa Girón, a flooded cave that is good for snorkelling. The deepest is Cueva Cuba Hungria at 470 m (1540 feet) in Sierra de Guamuhaya.
- The largest systems are Sistema Cavernario Palmarito, with 54 km (34 miles) of extensive caverns and galleries, and Cuevo del Santo Tomás, which extends for 46 km (29 miles). Both are near Viñales and can be visited by prior arrangement – contact the Sociedad Espeleológica de Cuba in the first instance.
- One of the best sites to visit is Cuevas de Bellamar. a couple of hours east of Havana. Its wonderful variety of features includes helictites, strange spaghetti-like shapes formed only under relatively rare conditions by capillary action.

Diving and snorkelling

- Jardines de la Reina or 'Gardens of the Queen', an archipelago of coral cays, mangrove and reef off the south coast, offers excellent visibility, mild currents and pristine reef. Along the whole 160-km (100-mile) length of the 'gardens' there is only one permanent floating hotel, La Tortuga Lodge; much of the diving is from live-aboard boats, which can be booked in advance through specialist local companies, such as Avalon Fishing and Diving Center.
- The best time to see whale sharks off the south coast of Cuba is between August and December. Generally, diving visibility is best in the south, and in the dry season or winter months, November to April.
- Among the many other sites are:
  - María la Gorda, 300 km (200 miles) southwest of Havana in Bahía de Corrientes
  - Isla de la Juventud, an island off the south coast
  - Varadero, to the east of Havana
  - Cienfuegos, on the southern coast, 250 km (150 miles) east of Havana
- There are specialist cave dive sites at El Brinco Cave, near Playa Girón (Bay of Pigs), and Saturno Cave, near Varadero.
- In addition there are numerous wrecks around the coast. One of the best known is the *Mortera*, a Spanish ship that sank in 1896 and is now known for its bull sharks. The *Cristobal Colón* is a nineteenth-century Spanish wreck near Santiago de Cuba.

PLACES TO VISIT

- **Ciénaga de Zapata** is the largest swamp in the Caribbean. It is easily accessible and can be enjoyed from boats and cars. There is a crocodile farm and visitor information centre on the main road through the reserve. See also Bird-watching section, p.202.
- **Old Havana.** No visit to Cuba is complete without some time spent in Havana. There is a lot to see – museums, churches, monasteries, even a camera obscura that gives a 360-degree perspective on the city. But it's the music-filled bars and cafes, the architecture of the squares and winding streets, the little coco taxis and 1950s cars that give Havana its unique character.
- **Trinidad.** A beautifully restored colonial town and World Heritage Site, Trinidad is a colourful combination of houses, churches, squares, narrow cobbled streets with bicycles and horse carriages.

WHERE TO STAY

Havana, Santiago de Cuba, Varadero and Trinidad all have a good choice of accommodation options, and there are a few good hotels around Isla de la Juventud.

FIND OUT MORE

General information, bird-watching and hiking: www.cuba.com

Cuban Embassy (UK)
Tel: 020 7240 2488
Fax: 020 7836 2602
www.cuba.embassyhomepage.com

Cubanacán (a commercial company handling 48 per cent of visitors to Cuba)
Tel: 020 7536 8176
Fax: 020 7537 7747
www.cubanacan.co.uk
www.cuba.tc/cubanacan.htm

Responsible Travel
www.responsibletravel.com

Climbing: Although there is no national climbing club, groups of climbers exist in Havana and Vinaless – see www.cubaclimbing.com

Caving: Sociedad Espeleológica de Cuba
Fax: (53-45) 24 2413 or (53-7) 881 5802

Diving: Avalon Diving Centre, www.divingincuba.com

Getting around: www.cubatravel.cu

R AT R A T S

A MA S A S

Diving is a main attractions, and with over 250 sites on Grand Cayman alone, there is something for everyone at any level. Don't overlook the less obvious reefs. A closer inspection of the coral reef will reveal many colourful smaller invertebrates. Public transport is somewhat elusive, and renting a car is the best way to travel independently (a local driving permit can be bought at the rental agency). Ferries and small planes ensure that the Caymans enjoy good accessibility.

ACTIVITIES

Diving and snorkelling

- An Official Snorkelling Guide available from the Cayman Islands Department of Transport lists the 39 best snorkelling sites on all three islands, of which 18 are on Grand Cayman. With visibility up to 30 metres (100 feet) you are bound to see spectacular marine life.

- The point where the shallow reefs end is the real reason many divers come here. Where the reef meets the open ocean the sea floor drops away dramatically to reveal spectacular drop-offs festooned with life – in particular the soft corals that reach out from the cliff face and giant sponges like the spectacular orange elephant ear.
- Cayman Brac is good for tunnels and swim-throughs which are found leading from the reef to the wall.
- Little Cayman has some of the most dramatic drop-offs, such as Bloody Bay Wall, which plummets over 1800 metres (6000 feet).
- The recent popularity of wreck diving has meant that many ships have been deliberately sent to the bottom, to supplement those already there. They attract a great deal of fish life and eventually become encrusted with corals. The most unusual must be the MV *Captain Keith Tibbett*, a 100-metre (330-foot) Russian destroyer purchased and sunk in 1996 just off the coast of Cayman Brac.
- Stingray City, where large numbers of rays gather waiting to be fed, is the experience of a lifetime and can be enjoyed both by divers and by snorkellers on the nearby sand bar. The rays, used to human interaction, are not shy and will brush up against visitors. One vital thing to learn is the stingray shuffle – a way of walking that reduces the chances of accidentally stepping on one.

Bird-watching

- Cayman Brac, at just 21 km (13 miles) long and 1.5 km (1 mile) wide, has recorded nearly 200 resident or migrating species. In April and May you may well see over 50 species in a single day – keep an eye out for the endangered Cayman Brac parrot if you visit the Brac Parrot Reserve.
- Little Cayman's many swamps and lagoons make it a favourite nesting ground for red-footed boobies and frigatebirds. Some 5000 breeding pairs of boobies can be seen at the Booby Pond Nature Reserve. The island is also a good place to look for the threatened West Indian whistling duck.
- On Grand Cayman the Governor Michael Gore Bird Sanctuary is a wetland reserve with around 60 species of birds.

Walking and hiking

- The Mastic Trail, a guided 3-km (2-mile) nature trail on Grand Cayman, traverses the island through farmland, woodland and mangroves and takes about 2½ hours.
- Cayman Brac has about 13 km (8 miles) of footpaths and trails with interpretive signs providing information about the birds and reptiles of the island. It also has over 100 caves and some challenging sea cliffs for climbing.

Turtle-watching

- Seven Mile Beach, just outside the capital George Town on Grand Cayman, is a great place for an early morning walk. If you're lucky you could have it to yourself or you might bump into members of the Department of the Environment who are out every morning from May to October looking for tracks left by the green turtles that use this beach for nesting. You can report any nesting or hatching activity you see to the Department of the Environment.

PLACES TO VISIT

- **Cayman Turtle Farm.** If you fail to spot a turtle while snorkelling or diving, a trip to the turtle farm will ensure you see them at all stages of development. From 10-cm (4-inch) long hatchlings to the grand 1-metre (3-foot) long adults there are hundreds of turtles to be seen. The farm is currently undergoing a major redevelopment to become a marine theme park that will include an interactive turtle area, snorkel lagoon, predator tank and aviary – the latest news on its reopening can be found at www.turtle.ky.
- **Queen Elizabeth II Botanic Park.** Inland on Grand Cayman, this park offers some 23 ha (60 acres) of botanic displays. On the 35-minute woodland trail you can see cactuses, palms, air plants and orchids. Over 200 species have been recorded in the park, and it is also the site of a breeding programme for the Cayman blue iguana.
- **Hell.** An unusual formation of jagged, blackened rock said to resemble the charred remains of hell fire. Most people visit to send a postcard from Hell from the local sub-post office.
- **Pedro St James.** The oldest building on Grand Cayman and the birthplace of democracy in the Cayman Islands. The first government was formed here in 1831 and just four years later the Slavery Abolition Act was read.
- **Pirate Week.** A wild week at the end of October centred around George Town harbour, this includes parades, regattas, cultural and historical days and treasure hunts.

WHERE TO STAY

Grand Cayman receives 99 per cent of tourists – George Town is particularly busy, with incoming cruise ship passengers peaking at over 20,000 per day. For an escape, and to take advantage of the best scuba diving, explore the options on Cayman Brac and Little Cayman. Hotels tend to be on the expensive side, but there is a good choice, even on the smaller islands.

FIND OUT MORE

Cayman Islands Department of Tourism (UK)
Tel: 020 7491 7771
Fax: 020 7409 7773
www.caymanislands.co.uk

National Trust:
www.nationaltrust.org.ky

Bird-watching, diving, events and planning: www.naturecayman.com

Diving: www.divecayman.ky

Information about Little Cayman and Cayman Brac: www.naturecayman.com

Getting around: www.cayman.org

JAMAICA

Dense vegetation means that this island is a haven for flora and fauna, especially flowering plants, with over 3,000 species including bougainvillea and orchids. The Blue Mountains are one of the most famous areas, thanks to the perfect conditions for coffee production. The Blue and John Crow

Mountains National Park, a mountain rainforest situated in the island's east, is also popular. Jamaica has diverse bird life typical of the Caribbean. There are around 200 species, over 25 of which are endemic. These include parrots, todies, cuckoos, pelicans and hummingbirds. The only native land mammal is the endangered Jamaican hutia, or cony, a guinea-pig-like large, brown rodent.

Find out more: www.jamaicans.com

PUERTO RICO

There's huge diversity within the natural attractions of Puerto Rico. El Yunque rainforest has the highest number of visitors of any natural site on the island, thanks mainly to the 240 species of trees. There are 350 species of birds on the island, as well as indigenous creatures such as tree-toads. Accessibility is generally good, but driving can be a frustrating experience. The coastal highways are the most scenic, but are congested. Some of the roads in the mountains are too narrow for cars, so check with the rental agency before you set off. Puerto Rico is a territory of the United States, though has a degree of autonomy and its own political system.

ACTIVITIES

Hiking and walking

- The Caribbean National Rainforest, known locally as El Yunque, is perhaps the most popular and best known of Puerto Rico's forest reserves. Situated in the east of the island, it has the interesting distinction of being the only tropical rainforest in the US National Forest system.
- Hiking in El Yunque is one of the safest ways to visit a tropical rainforest. There are more than 50 species of orchid to look out for, magnificent giant ferns bordering many of the paths and tabonuco trees oozing a medicinal-smelling sap. Here you will also hear one of the most characteristic sounds of Puerto Rico – the distinctive 'ko-kee' call of the coqui frog. There is a visitors' centre from which you can get details and free maps of the various trails.
- Only two hours' drive from the damp rainforest of El Yunque, the Guánica Biosphere Reserve, in the southwest of the island, is 3500 ha (9000 acres) of tropical dry forest. There are plenty of trails and some beautiful unspoilt coastline to be explored.

Bird-watching

- El Yunque is home to over 60 species of birds. The most famous of these is the Puerto Rican parrot, which you are unlikely to see. Much more common is the Puerto Rican tody.
- In the dry forest of Guánica there are over 130 bird species, including the endangered Puerto Rican nightjar.
- Along the coast pelicans can be seen diving into the water hunting fish, or gliding effortlessly just centimetres above the surface of the sea.

Diving and snorkelling

Puerto Rico is surrounded by reefs but perhaps the best diving is to the east of the island around the smaller islands just offshore.

- The La Cordillera islands are very popular, with beautiful coral gardens.
- The Vieques Channel has extensive reefs and good drop-offs.
- The island of Culebra is good for snorkelling.

Bioluminescence

- Two of the best locations for witnessing this phenomenon – the natural light show that occurs when the plankton in the water give off light when disturbed – are Mosquito Bay on Vieques and Phosphorescent Bay at La Parguera. Snorkelling at night, you can watch your finning feet leave a trail of blue-green sparks in their wake.

Caving

- Puerto Rico boasts the third-largest cave network in the world, beneath the rugged karst country in the north of the island. Many of the caves are home to huge bat roosts from which thousands of bats emerge to feed on insects. On dark nights Puerto Rican boas can sometimes be seen hanging from the trees at the caves' entrance trying to snatch emerging bats.
- The Cavernas del Río Camuy spread over 16 km (10 miles) and have 17 entrances. The Parque de las Cavernas del Río Camuy has a visitors' centre and trips into the caves. If you want to explore further there are several organizations that arrange caving adventures for all levels of cavers.

PLACES TO VISIT

- **Mona Island.** For the more serious adventurer, this little island to the southwest of Puerto Rico has the dual attractions of being home to the huge Mona iguana and being surrounded by large areas of healthy coral. You may see elkhorn corals and groupers, but the best sight has to be one of the estimated thousand hawksbill turtles that cruise these waters.
- **Old San Juan.** This World Heritage Site is an architectural delight. Strolling around the narrow streets, soaking up over 500 years of Spanish colonial history, is a great way to spend a day. The town can become crowded due to the frequency of large cruise ships in the port, but it is full of historic churches, museums and stunning town houses; it also has some of the best restaurants and shopping on the island.
- **Fuerte San Felipe del Morro**. This sixteenth-century fort is impressive for both its architecture and its dramatic setting overlooking the Atlantic and the port of San Juan.
- **Arecibo Observatory.** Home to the world's largest radio telescope, made famous by James Bond. The almost 8-ha (20-acre) dish is built in a natural sinkhole and surrounded by hills – an unusual sight in this natural setting.

WHERE TO STAY

Hotel classification is erratic, so check carefully when booking. Guesthouses are very popular, and can be booked through the Tourism Company, which can also help book paradores, private country inns offering good value.

FIND OUT MORE

Puerto Rico Tourism Company (Puerto Rico):
Tel: (787) 721-2400
www.gotopuertorico.com

Caving in Camuy:
www.camuypuertorico.com

Diving: www.puertoricodiving.com

Hiking:
Bano de Oro trail:
www.hechoenpuertorico.org/yunque/hik_car6.htm
USDA Forest service: www.fs.fed.us
El Yunque: www.iit.edu/~PR/elyunque

Bird-watching:
www.prwow.com/html/birdswatching
www.elyunque.com/birds

For more information on bird-watching in Puerto Rico, you can also call the Conservation Trust Office in San Juan: (787) 722 5834.

AR S A S

BR T SH R S A S

A collection of over 30 islands and cays, around 16 of which are inhabited. The largest are Tortola, Virgin Gorda and Anegada. Of the uninhabited islands, West Dog, Great Tobago and Diamond Cay all have bird sanctuaries that are accessible to visitors by boat. The BVI do not have any endemic birds of their own, but you can see species endemic to the Caribbean, as well as the endangered West Indian whistling duck and the piping plover. The Anegada rock iguana is found only on that island. The BVI are semi-tropical and prevalent trees include the wild tamarind and the indigenous white cedar, the national tree.

Find out more: www.bvitourism.com

A T UA & BARBU A

The intriguing island of Barbuda was once a trove of sunken ships and now boasts treasure of a different kind in the form of the Frigate Bird Sanctuary, situated in the Codrington Lagoon. There are very few predators here and this nesting site is one of the most important in the world. Barbuda is a flat coral island with an area of 175 sq km (68 sq miles). The roads are reasonable, and car hire is easy to arrange. There are a few luxurious resorts, but Barbuda remains a relatively undeveloped island. Activities here are suitably relaxed, and include beachcombing (on the northeastern Atlantic coast), fishing and hunting.

ACTIVITIES

Bird-watching

A highlight for any nature lover, Codrington Lagoon has some of the most remote and unspoilt tracts of mangrove of any Caribbean island and its bird sanctuary has one of the world's largest nesting colonies of magnificent frigatebirds – literally thousands of birds. By October the adult males have returned to the reserve and begun their mating displays; the best time to see chicks of different ages is January.

Diving

Both islands boast warm waters with excellent underwater visibility – over 40 metres (140 feet) is not unusual. Conditions are ideal for both diving and snorkelling, though diving facilities are better in Antigua. Part of one of the best-known sites, Cades Reef, is protected in an underwater park; many others are no more than 40 minutes offshore. Many as yet unexplored wrecks can be found off both islands.

Beaches

One of Barbuda's claims to fame is its amazing sandy beaches, some of which have a slight pinkish hue. The most accessible of them stretch for many kilometres along the western (leeward) shore, facing the Caribbean Sea, and have calm, clear blue waters. The Atlantic coast has a raw beauty with limestone terraces and even more deserted beaches that are accessible on foot, though only some are designated swimming beaches.

WHERE TO STAY

There is a good range of hotels and accommodation on both islands; on Barbuda, there are several guesthouses and cottages available to rent in the village of Codrington.

FIND OUT MORE

Antigua & Barbuda Department of Tourism (Antigua)
Tel: 268 462 0480
Fax: 268 462 2483
Email: deptourism@antigua.gov.ag
www.ab.gov.ag

Antigua & Barbuda Tourist Board (UK)
Tel: 020 7258 0070
Fax: 020 7258 3826
www.antigua-barbuda.com
www.antigua-barbuda.org

Codrington:
www.antiguamuseums.org/barbuda.htm

Diving: www.antigua-barbuda.org/Agdive01
www.divetravel.netfirms.com

Hiking & bird-watching:
www.antigua-barbuda.org/Agact01.htm#hiking

www.antigua-barbuda.com/travel_tourism/sports_activities/activities

www.ab.gov.ag/gov_v2/visitors/activities

Getting around: www.antigua-barbuda.org/Agtsp01

M TS RRAT

Montserrat was once another idyllic island along the Caribbean chain, its claims to fame being forested scenery, waterfalls, the endemic Montserrat oriole, the endangered 'mountain chicken' (a species of frog) and the pretty town of Plymouth. In July 1995 that all changed when steam and ash started to vent from the Soufrière Hills volcano; eruptions have continued on and off ever since. Although tourist activity is now limited to the 'safe zone' in the central and northern parts, much of the island still has forested hills and the coast is dotted with black volcanic-sand beaches.

ACTIVITIES

Volcano-watching

- Visits to the Montserrat Volcano Observatory offer a great view of the volcano and scientists can demonstrate how the continuing volcanic activity is monitored.

Walking and hiking

- This is obviously more limited than it used to be, but there are still places to explore in the Centre Hills and Silver Hills, including the Oriole Walkway. There are also walks around the coast to see the black and white sand beaches, like the trail from Little Bay to Rendezvous Bay. This offers views of the ruined town of Plymouth, and there has recently even been talk of running guided trips there, but check with the tourist board, as all information is liable to change.

Bird-watching

In the forests of the 'safe zone' the endangered Montserrat oriole is the prize for keen birders, though its range has been radically reduced by the destruction of the forests around the crater.

Diving

There is one operator and a range of dive sites north of the Volcanic Maritime Exclusion Zone. One site includes surfacing in a bat cave!

WHERE TO STAY

Mostly guesthouses and villas that can be rented privately. There are only two hotels on the island. Possibly best combined with a trip to Antigua, St-Martin or St Kitts and Nevis.

FIND OUT MORE

Montserrat Tourist Board (Montserrat)
Tel: (664) 491 2230/8730
Fax: (664) 491 7430
E-mail: info@montserrattourism.ms
Website: www.visitmontserrat.com

Montserrat Volcano Observatory
www.mvo.ms / tel (664) 4915647

Hiking and bird-watching:
The Montserrat National Trust (MNT)
Tel: (664) 491 3086
Email: mnatrust@candw.ms
www.montserratnationaltrust.com

Diving: www.montserratdiving.com

Getting around:
www.visitmontserrat.com/index.php?categoryid=29

WINDWARD ISLANDS

DOMINICA

Dominica has some of best rainforest in the Caribbean – 60 per cent of the island is covered by it. Most of the country is mountainous, with many areas plunging steeply to the sea. The rainforest-covered Morne Trois Pitons National Park in the interior is a UNESCO World Heritage Site. Other major parks include Cabrits, an eighteenth-century British fort. Dominica has well-maintained roads, a good bus service and efficient taxis.

ACTIVITIES

Walking and hiking

- There are hiking trails throughout the island to suit all abilities.
- The toughest and most rewarding are in the Morne Trois Pitons National Park. The strenuous full-day walk to Boiling Lake and the Valley of Desolation involves a three-hour climb during which you emerge from lush rainforest on to mountain ridges with fine views. Check out the level of activity of Boiling Lake before you go and take an experienced guide with you for safety. The viewpoint is above the lake and the guides recommend that you do not spend more than a few minutes so close to the sulphurous fumes. In the volcanic landscape of the Valley of Desolation, water bubbles at scalding temperatures and the stream bed is coated with brilliantly coloured mineral deposits.
- Another rewarding hike is to Middleham Falls, about 45 minutes each way from Laudat or Cochrane.
- Shorter trails a few minutes into the rainforest just outside Roseau include those to Trafalgar Falls and Emerald Pool, which are also beautiful. Look out for freshwater crabs on the trail, especially after rain.
- There are also gentle trails through the rainforest in the Syndicate area of the Morne Diablotin National Park.
- Dominica's mountainous terrain means that waterfalls abound. On the edge of Morne Trois Pitons is a variety of falls, including the easily accessible Middleham Falls, Trafalgar Falls and Emerald Pool. To the southeast are the stunning but relatively inaccessible Sari Sari Falls and Victoria Falls.

Bird-watching

- The best place to look for both of Dominica's parrots – the imperial parrot or sisserou and the red-necked parrot or jaco – is in the Syndicate rainforest. A number of species of hummingbirds can also be seen there, although some of the best sightings can be in the gardens of guesthouses and hotels where these colourful little birds are drawn to feeders. Antillean euphonies, lesser Antillean pewees, bananaquits, rufous-throated solitaires and plumbeous warblers are also common in the forest.
- The swamps of Cabrits and Glanvillia are important localities for waterbirds – crakes, rails, herons and egrets.

Turtle-watching

- From March to September giant leatherback, hawksbill and green turtles use the Atlantic coast at Rosalie Bay to lay their eggs. Check out the Rosalie Sea Turtle Initiative (RoSTI) protection programme to obtain details about levels of activity.

Whale-watching

- The Caribbean shores of Dominica are some of the best places in the world to see sperm whales. They are resident all year round but the ideal time for viewing is between November and April, when the seas are calm and females and young are being courted by the males, which creates a great deal of social activity. Whale-watching tours depart from Roseau.

Diving and snorkelling

- Along the southwest coast, Champagne Pool is a snorkelling spot where volcanic gas bubbles rise in columns from the seabed.
- The best locations for scuba diving are within the Soufrière-Scotts Head Marine Reserve (SSMR), where the volcanic landscape continues underwater with marine pinnacles and drop-offs.

- There are other interesting dive sites within the marine area of Cabrits National Park.

Rainforest-viewing

- The aerial tramway near Titou Gorge is a pleasant way to experience the rainforest, especially the canopy, complete with bromeliads. On the way up there is good view of the Breakfast River valley and it is possible to get off at the top, hike through the forest and rejoin the tramway on the way down. As well as birds and butterflies, agoutis can sometimes be seen on the forest floor beneath.

Volcanic lakes and streams

- In addition to Boiling Lake (see under Walking and hiking, p.207e), there are other crater lakes that are no longer active but still make pleasant excursions, such as Boeri Lake and Freshwater Lake. A visitor reception centre was recently opened at Freshwater Lake and a local community group offers recreational canoe tours and kayaking.
- For those wanting a taste of the volcanic activity without the hard climb, the bubbling sulphur springs of Wotten Waven are just a few metres from the road. Soufrière Sulphur Springs, in the southern part of the island, are a short, gentle hike. It's worth noting that Soufrière, like all the volcanic features, is always changing: sometimes there is little steam emerging from the ground, but the yellow sulphur rocks are still attractive.

PLACES TO VISIT

- **Indian River Swamp.** Lying just south of Portsmouth in one of the few areas of flat land on the island, the swamp is worth a canoe trip to explore the contorted roots of the pterocarpus or bloodwood trees, and to watch the crabs and shorebirds.
- **Roseau**. The capital of the island, where the botanic garden, founded in 1890, has a collection of plants and some captive Dominican parrots.
- **Soufrière and Scotts Head.** Brightly painted fishing boats and a pretty church make this village one of the most picturesque on the island.
- **Fort Shirley.** An eighteenth-century British military garrison in Cabrits National Park, north of Portsmouth.
- **Carib settlements.** On the Atlantic (eastern) side of the island, north of Castle Bruce, lies the Carib Territory. Handicrafts are on sale from thatched huts along the road. A recently opened Carib cultural village named Kalinago Barana Auté or 'Carib village by the sea' depicts the traditional architecture and lifestyle of the Carib people.

WHERE TO STAY

Accommodation is mainly private guest houses, apartments, lodges and villas, though there is a good geographical spread of these across the island. Roseau has a couple of hotels. Local specialists such as Nature Island Destinations offer free impartial advice.

FIND OUT MORE

Tourist Board of Dominica (Dominica)
Tel: (767) 448 2045
Fax: (767) 448 5840
www.dominica.dm
www.avirtualdominica.com
www.lennoxhonychurch.com
www.dhta.org

Nature Island Destinations
www.natureisland.com

Hiking:
www.dominica.dm
www.rainforestrams.com

Whale watching:
www.anchoragehotel.dm
www.castlecomfortdivelodge.dm
www.natureislanddive.com

Bird-watching:
www.natureisland.com/bird+bot

Diving: www.marinereserves.com

Turtle-watching: Rosalie Sea Turtle Initiative or RoSTI

www.marinecreatures.com

Getting around:
www.natureisland.com/Car

GRENADA

One-sixth of Grenada is protected as either parks or wildlife sanctuaries, and the tropical rainforests are home to a variety of wildlife and waterfalls. The most popular of the parks is the Grand Etang Forest Reserve, high up in the mountains of the interior. The forest provides shelter for a host of the island's many species of birds: the broad-winged hawk, Lesser Antillean swift, Antillean euphonia, purple-throated carib, Antillean crested hummingbird and the Lesser Antillean tanager are all common sights. The hiking trails usually take in the Seven Sisters Falls and Mount Qua Qua.

ACTIVITIES

Walking and hiking

- The Grand Etang National Park Forest Reserve, like the rest of the island, was hit heavily by Hurricane Ivan in 2004, and seeing a normally lush evergreen forest stripped bare after a hurricane has passed by is a rare experience. However, it is also extraordinary to see how quickly the forest recovers: at the time fallen trees closed many of the numerous hiking trails, but these are now opening up again. They vary in difficulty from the short (15-minute) Morne La Baye Nature Trail to the often wet and muddy, and much longer, Mount Qua Qua trail.

Waterfalls

- Of the many found throughout the island, the picturesque Concord Falls are perhaps the best known and most easily accessible by road. They are often used as a place to have a cooling dip, though care should be taken, especially after heavy rainfall.
- Seven Sisters Falls in the Grand Etang National Park are a series of seven waterfalls that requires a 30-minute trek to the first.
- The recently discovered Victoria Falls, in a beautifu,l unspoilt location in the foothills of Mount St Catherine are also accessible by foot.

Diving

- Grenada is a great diving destination with over 28 sites including reefs, wrecks and walls. The majority are within 20 minutes of the dive shops and offer the chance to see beautiful coral gardens as well as angel fish, tang, parrotfish, moray eels and lobsters. Popular sites include Boss Reef, Valley of Whales and Flamingo Bay. For beginners Spice Island Reef is recommended.

- Sandy Island, a 'classic' Caribbean island with palm-fringed white sandy beaches, is a favourite snorkelling and diving spot just off Carriacou.
- Grenada also boasts the biggest wreck in the Caribbean – *Bianca C*, a 180-metre (600-foot) cruise ship that sank in 1961 and now lies in 50 metres (165 feet) of water near Whibble Reef. This advanced dive should be undertaken only by experienced divers accompanied by a dive operator.
- On Molinere Reef a more accessible 13-metre (42-foot) sloop, the *Buccaneer*, can be explored and at Red Buoy are the coral-encrusted remains of an eighteenth-century wreck.

Bird-watching

- One of the best places for birders is the national park around Levera Pond. With its large mangrove system and a bird sanctuary, it's good for waders, shore birds and the endangered hook-billed kite.
- Try looking for the rare Grenada dove in the scrubby areas of Mount Hartman National Park.
- In the rainforest there are emerald-throated hummingbirds and red-necked pigeons.

PLACES TO VISIT

- **Spice factories.** Grenada is the world's second-largest producer of nutmeg. At Gouyave and Grenville nutmeg-processing stations, tours let you watch the workers sorting, grading and de-husking the nutmeg, then bagging up the spice and preparing it for export. The processes haven't changed much in centuries – and it smells wonderful.
- **St George's.** Grenada's capital has one of the prettiest harbours in the Caribbean, a perfect horseshoe shape. A historic walking tour that is well worth doing points out the buildings of interest. The architecture is a lovely mix of French provincial and English Georgian styles.
- **Fort George.** Built in 1706, the fort offers a spectacular view of the town and harbour. It was at the centre of the 1983 political disturbance during which Prime Minister Maurice Bishop and other members of the government were executed by a group from the People's Revolutionary Government. Look out for evidence of the fighting in the walls of the fort.
- **Market Square.** Enjoy the hustle and bustle of a true Caribbean market. This is where much of the island's produce is bought and sold. It's a noisy, colourful place where you'll rub shoulders with the local people.
- **Mount Rich Amerindians petroglyphs.** The only extensive area of carvings left on the island, depicting the way of life of the earliest inhabitants. The carvings can be seen on the tops and sides of the large stones on the riverbank.

WHERE TO STAY

St George's and the beach areas of Grand Anse are in the southwest of the island, where most of the accommodation is centred. Properties range from luxury all-inclusive resorts to private guesthouses. On the southern coast, La Sagesse Bay is more secluded.

FIND OUT MORE

Grenada Board of Tourism (Grenada)
Tel: (473) 440 2279
Fax: (473) 440 6637
Email: gbt@spiceisle.com
www.grenadagrenadines.com

Diving:
www.grenadaexplorer.com/diving
www.divegrenada.com

Bird-watching:
www.grenadagrenadines.com
www.birdingpal.org/Grenada

Getting around: www.grenada-guide.info/getting.around

MARTINIQUE

Martinique is at a three-way crossroads of major ecosystems – coral reefs, mangroves and seagrass beds. This results in outstanding ecological and scientific phenomena around its coastal areas. The island is notable for its large, indented bays where extensive seagrass beds and mangrove forests are to be found. The island has some 30 diving clubs and 2000 licensed divers. The most popular diving area is around Cap Salomon, south of Fort-de-France Bay.

Find out more: www.martinique.org

ST LUCIA

St Lucia's beaches are among the finest in the Caribbean; however, the centre of the island is dominated by a range of hills and mountains, much of it covered in rainforest. Morne Fortune – the 'hill of good luck' – has panoramic views across the capital of Castries. The most impressive park is the 7340-ha (19,000-acre) National Forest Reserve, although Pigeon Island, just north of Gros Islet, also has some excellent hiking. Most of the major centres are covered by a good road network, and car hire, buses and taxis are available.

ACTIVITIES

Walking and hiking

There are three main trails into forestry land:

- The Edmund Forest Reserve trail begins inland from Soufrière and traverses tropical rainforest with bromeliads, vines and orchids. The Des Cartiers Rainforest trail at the Quilesse Forest Reserve is home to the endangered St Lucia parrot. Both these trails take a few hours and can be very wet underfoot at certain times of the year – the driest season is from January to April.
- The Barre de L'Isle trail, also through rainforest, leads to the top of Morne la Combe and takes about three hours to complete.
- It is possible to hike to the top of Gros Piton, but an expert guide is required to attempt this – you can hire one for a nominal fee at the

Interpretive Centre in the nearby town of Fond Gens Libre.

Bird-watching

- Many beautiful birds can be seen around the gardens of hotels and in the botanical gardens. The island's most famous bird, the St Lucian parrot, is really hard to see, although you may be lucky if you are hiking in the Quilesse Forest Reserve.
- The Frigate Island Nature Reserve, a couple of islands on the Atlantic coast, can be visited through the St Lucia National Trust or local tour companies. In summer it is the breeding site for magnificent frigatebirds and several rare indigenous species such as the St Lucia oriole and Ramier pigeon.
- The Maria Islands Nature Reserve, east of Vieux Fort, is a sanctuary for noddies, terns and other seabirds, and can be visited through the St Lucia National Trust. It is also the only place to find two species of endemic reptiles – the St Lucia whiptail or ground lizard and the rarely seen kouwes snake.

Diving and snorkelling

The mountainous landscape of St Lucia continues underwater and makes for a variety of interesting dive sites – caves, drop-offs and pinnacles.

- Anse Chastanet, next to the Soufrière Marine Reserve and close to the Pitons, offers a variety of diving and snorkelling sites close to shore, or further up the coast by boat; those close to the Pitons have the bonus of stunning views of the mountains from the sea. Two of the best-known sites are the Key Hole Pinnacles – coral-covered seamounts – and *Lesleen M*, a freighter sunk to create an artificial reef at Anse Cochon.
- Other dive shops are based further north at Castries and Marigot Bay.

Volcanic landscapes and sulphur springs

- Views of the towering steep-sided peaks of the Pitons can be enjoyed from a boat, from the road, hiking or by helicopter on a scenic flight.
- The nearby Sulphur Springs are dramatic in a different way, with the smell of rotten eggs given off by hydrogen sulphide, the sound of bubbling mud and the moon-like rock steaming and hissing. You can get quite close to all this activity from the safety of walkways.

Beaches

There are small beaches and bays on the Caribbean shore to the southwest of the island and a long white-sand beach in the northwest, connecting Gros Islet and Pigeon Point. Reduit Beach and Choc Beach are also popular spots near the tourist resorts on the northern tip of the island.

PLACES TO VISIT

- **Diamond Botanical Gardens.** On the outskirts of the coastal town of Soufrière, this is a pretty spot to see tropical trees, flowers and a variety of birds. There is also a small waterfall with orange-coloured mineral-stained rocks and a mineral bath based on the hot, volcanic springs.
- **Marigot Bay.** A sheltered anchorage for yachts with a palm-fringed beach.
- **Pigeon Island National Park.** Pigeon Island was fortified by a British admiral in the eighteenth century to keep watch over the French fleet on Martinique. The island became more accessible in the 1970s with the addition of a causeway. The ruins of the fortress at Fort Rodney Hill are now administered by the St Lucia National Trust.

WHERE TO STAY

St Lucia offers a typically Caribbean choice of accommodation, from huge all-inclusive resorts to luxury villas, small inns and guesthouses. Most is centred around the north and south of the island.

FIND OUT MORE

St Lucia Tourist Board (St Lucia)
Tel: 758 452 40 94
Fax: 758 453 11 21
www.stlucia.org
www.sluonestop.com
www.stlucia.com

St Lucia National Trust:
www.slunatrust.org

Forests department: www.slumaffe.org

Nature guide: www.geographia.com/st-lucia/

Diving: www.stlucia.org/activity/diving

Hiking:
www.visitslu.com/discover_slu/attractions/hiking

Getting around:
www.geographia.com/st-lucia

ST VINCENT AND THE GRENADINES

The Grenadines consist of some 32 islands and cays stretching south from St Vincent, only eight of which are populated. The shallow waters that surround them are a haven – coral reefs fringe the islands and extend up to beds of sea grass that house lobsters, conch and turtles, which can be found in such areas as the Tobago Cays, to the north of Prune (Palm) Island. Whales are frequently sighted off Petit Nevis, and large iguanas can be found on some of the waterless rocks and cays. Among the many birds found in St Vincent are the Caribbean eleania, the trembler, the bananaquit and the Antillean crested hummingbird. There are a number of protected areas, such as Tobago Cays National Park and Sandy Island.

Find out more: www.svgtourism.com

SOUTHERN CARIBBEAN

TRINIDAD AND TOBAGO

Trinidad and Tobago are among the Caribbean's top birding destinations, with about 420 species. Getting there and getting around are both easy – there are plenty of flights, plenty of buses and taxis, and a ferry between the two islands. Mountainous parts of both islands are covered by rainforest. The Tobago Forest Reserve, which covers the central highlands, was established in 1765, making it the oldest protected forest in the Caribbean. The offshore island of Little Tobago is worth a visit for its scenery and dramatic landscape, even if you are not a keen birder.

ACTIVITIES

Bird-watching

- From November to March, Trinidad has the wonderful evening spectacle, accessible by organized boat trip, of scarlet ibis flying in to their roosting sites in Caroni Swamp. Although the light may be low, and you are a little distant from the roosting trees, the sight of such brilliantly coloured birds in large numbers is a real treat. Sometimes egrets share a tree to make a dramatic red and white combination.
- Asa Wright Nature Centre can be visited as a day trip, but an overnight stay allows you to be up with the dawn chorus spotting some of the 460 species that fly around the veranda and grounds. These include the brilliantly coloured green honeycreeper, purple honeycreeper, violaceous euphonia, violaceous trogon, channel-billed toucan, crested oropendula and a variety of humming birds. It is also possible to see the golden-headed manakin and, at certain times of the year, the lekking display of the black and white manakin.
- On certain days, residents of Asa Wright can join a guided visit to Dunston Cave to see one of the few populations of oil birds left on the island.
- The route from Asa Wright to Blanchisseuse is also popular with birders, as many species can be seen along the edge of the road.
- Visits to Nariva Swamp National Park on the eastern side of the island and to Arima Reserve, a grassland reserve that lies between Asa Wright and Nariva, can be very rewarding. Just inland from the east coast road, next to Nariva Swamp, you may spot red-bellied macaws and yellow-headed parrots in the palms. Access to the swamps may be by boat or four-wheel-drive vehicle, depending on the season.
- The Pointe-a-Pierre Wildfowl Trust on the west coast, north of San Fernando allows you to see some of the birds up close. There are about 90 species here, including herons, waders and waterfowl. There is a gentle walk around a lake, some woodland and a chance to see scarlet ibis in captivity. The trust is a centre for environmental education and breeding endangered birds, and visits need to be arranged in advance.
- The protected forests in the interior of Tobago are home to blue-backed manakins, sabre-winged hummingbirds and ruby topazes, which, despite being tiny, are easily spotted with the help of a good guide.
- From Tobago a boat trip to Little Tobago is a must. It's a seabird sanctuary for red-billed tropicbirds, brown boobies, Audubon's shearwaters, brown noddies, sooty terns and magnificent frigatebirds. Watching them soar over the cliffs, then dive for fish, is a real treat.

Nature spotting

- On the way to see the scarlet ibis in Caroni Swamp, look out for the cute, although usually inanimate, fluffy ball of a sleeping silky anteater. This is the only island in the Caribbean where you can find them. Snakes are common in the swamp, as are the comical-looking anableps or four-eyed fish – so called as their two eyes are split so that they can see below as well as above the surface of the water.
- Nariva (see also under Bird-watching, above) is the best place to see Trinidad's two species of monkey, the white-throated capuchin and the red howler monkey. The howlers are more easily heard than seen and it can be frustrating trying to keep up with them as they venture deep into wet, inaccessible parts of the forest. Hiking is the only option and access varies with the seasons, so check with the tourist office.

Turtle-watching

- Trinidad is one of the world's best leatherback-turtle nesting sites. The turtle conservation programme on Matura Bay on the east coast is a good spot to see them in the nesting season between March and August, with April to June the peak time. Nature Seekers, a local conservation organization, co-ordinates the turtle-watching and can advise about permits.
- Other beaches along the north coast that are used by turtles are Grande Rivière Bay, Paria Bay, Madamas Bay and Grand Taciribe. Turtles may emerge from the water at any time of night, and noise and light can disturb them, so expect a long, quiet wait.
- Leatherback and green turtles also nest on Tobago's leeward beaches, such as Mount Irvine Bay.

Walking and hiking

- There are lots of walking trails in Trinidad, especially in the mountains in the north of the island, and along tracks near the north coast. Much is inaccessible by car, making it an attractive proposition for campers who want to visit the most remote areas.
- Mount St Benedict, the site of a Benedictine monastery, is surrounded by rainforest and is a haven for hikers and birders. A favourite walk at St Benedict is the two-hour Donkey Trail followed by refreshments at Pax Guest House.
- There are easy guided walks through the forests at Asa Wright (see Bird-watching, above), with the bonus of excellent birding from the trails. Day visitors should book ahead.
- Hiking trails are also a feature of the Tobago Forest Reserve. The best known is the Gilpin Trail, a 5-km (3-mile) walk through the jungle. Other trails lead to waterfalls and high ridges where there are views across the island to the coast. All are good for birding, and the local nature guides know the best trails for different species.

Diving and snorkelling

- The strong currents round Tobago make for great drift diving. Most of the best is at the northeastern end of the island. There are also a few good snorkelling spots for non-divers off Pigeon Point and Buccoo Reef in the southwest.

Beaches and bays

- Tobago has some charming, picture-postcard beaches: Englishman's Bay, Bloody Bay and Parltuvier are the epitome of the Caribbean idyll. The majority of the other white-sand beaches are at the southwestern end of the island, from Store Bay and Pigeon Point to Arnos Vale Bay. This is also where most tourist development has occurred.

- Trinidad has a few sandy beaches. Maracas, 40 minutes from Port of Spain, is one of the most popular. Some of the north-coast beaches are not great for swimming, but the scenery is worth the trip. The beaches on the east coast are exposed to the wild Atlantic wind and waves.

PLACES TO VISIT

TRINIDAD

- **Carnival in Port of Spain.** Although this is an event rather than a place, it is one of the most famous aspects of Trinidad, the largest and most spectacular carnival in the Caribbean and a riot of sound and colour. The date changes each year with Easter but it's always the two days preceding Ash Wednesday. The main spectacle is on the Tuesday when thousands of costumed 'band' members parade and dance through the streets.
- **The hot tar lake of La Brea.** This unusual geological feature is in the southwest of the island. A guide will take you out walking across this baking black desert with the texture of gooey toffee. Sometimes the corpse of an animal or bird is left sticking out of the surface, a reminder of what can happen if you stand for too long in the wrong place.

TOBAGO

- **Fort King George, Scarborough.** There is a history museum in the fort, with exhibits about life on the plantations and trade between the Caribbean and Europe. And the views along the coast are spectacular!
- **Charlotteville.** A charming fishing village nestled into the mountains just 4 km (2½ miles) from Speyside. From the village there are several walks through the countryside, including a 20-minute stroll to Pirate's Bay, which, if sea conditions are right, can be great for snorkelling.

WHERE TO STAY

In Trinidad, venturing south of Port of Spain means that your accommodation options become more limited, though there are still a few hotels and guesthouses. San Fernando and Mayaro are the best options. In Tobago, the western side has the bulk of the tourism, with large hotels. Heading east, past Scarborough, things are more rural, with much smaller inns and local restaurants. The Tourism Development Company has listings for all areas.

FIND OUT MORE

Trinidad and Tobago Tourist Office (UK)
Tel: 020 8350 1009
Fax: 020 8350 1011
www.visittnt.com

Tourism and Industrial Development Company of Trinidad and Tobago (TIDCO): www.tidco.co.tt

Environmental Management Authority: www.ema.co.tt

Asa Wright Nature Centre: www.asawright.org

Nature Seekers Incorporated turtle project: Tel: 868 667 9075, or e-mail natseek@tstt.net.tt

Pointe-a-Pierre Wildfowl Trust: www.petrotrin.com
Tel: 1(868) 658 4200, e-mail wildfowl.trust@petrotrin.com

Pax Guest House: www.paxguesthouse.com

Forestry: www.tobagowi.com/sites/forestresrv

Association of Tobago Diving Operators: www.tobagoscubadiving.com

Getting around: http://trinidad-guide.info/getting.around

ABC ISLANDS – ARUBA, BONAIRE AND CURAÇAO

One of the Caribbean's top diving destinations. All three islands have coral reef; Bonaire's is the most continuous and it is all a marine park. The most accessible part lies along the sheltered (westerly) side, but what is remarkable about it is that most of it lies only a few metres from the shore, with its drop-off within gentle swimming distance. All three islands have similar landscape and vegetation, which is characterized by one outstanding feature – it looks dry and almost desert-like in parts. Bonaire is quieter and less developed than the other two, but all are served by international airlines and connected by internal flights.

ACTIVITIES

Diving and snorkelling

ARUBA

- The wreck of the *Antilla* lies in shallow water close to shore and can be snorkelled as well as dived. New wrecks are being sunk all the time for the benefit of tourists.
- Patches of reef along the southwest lee shoreline offer good diving and snorkelling.
- For non-divers, a novel marine experience is a trip in the *Atlantis* submarine, which takes a 45-minute dive to depths of 45 metres (150 feet).

BONAIRE

- Bonaire has a number of dive resorts, many of which offer trips to sites further afield. The proximity to the shore of the reef and its drop-off makes them ideal for divers and snorkellers of all levels of experience. It also has remarkable biodiversity: some claim it has more fish species than any other Caribbean reef. It is not the place to spot sharks or other pelagic species, but rather a wonderful location to observe the abundance of invertebrates and brilliantly coloured fish.
- Many dives can be reached from the shore and a map with numbered sites allows you to be independent of dive operators, as long as you have transport.
- There are also numerous dive sites further out, with moorings for boats, to explore the drop-off and the wreck of the *Hilma Hooker* freighter.
- The nearby island of Klein Bonaire is an attractive diving and snorkelling spot that can be reached by boat within a few minutes of the main island.

CURAÇAO

- An underwater park along the southeast coast can be enjoyed by diving, snorkelling or through a glass-bottomed boat.

Bird-watching

- The yellow-shouldered Amazon parrot or lora is endemic to the islands, but the only breeding population left is on Bonaire. If you are lucky you may see them feeding on cactus fruit.
- On Aruba the key place for birders is Arikok National Park, where the Caribbean parakeet roosts and is easily spotted. Arikok is good for a variety of other birds, too, including orioles, yellow warblers and caracaras, which hang around the large granite boulders. The Aruban burrowing owl or shoco may be seen very early in the morning perched on limestone rocks. You are also likely to see whip-tailed and anole lizards in this park.
- On Bonaire the local race of parakeets can be found in Washington-Slagbaai National Park and at other locations around the island.
- In the dry season a good way to see a variety of birds in a relatively short time is to wait by one of the two freshwater springs in Washington-Slagbaai. Pos Mangel and Put Bronswinkel attract mockingbirds, ground doves, yellow warblers, bananaquits and black-faced grassquits, among others. Green iguanas may also be seen around Pos Mangel.
- Washington-Slagbaai is also the place to see Caribbean flamingos feeding (check their whereabouts at the park gate), but the salt ponds in the south of the island are where they go to nest. Even with the most powerful binoculars it is hard to see more than a thin strip of pink, as they are concentrated in a restricted area owned by the salt works.
- Curaçao is more developed than the other two islands, but has a protected area around St Christoffel mountain, which is a haven for wildlife similar to that found on Aruba and Bonaire.

Turtle-watching

- Turtles come to nest in Bonaire between the months of May and July: contact Sea Turtle Conservation Bonaire for details.

Kayaking

- The Lac Bay mangrove forest is one of Bonaire's protected areas; kayak tours with a naturalist guide based at the Mangrove Information and Kayak Center are the dry way to see baby reef fish hiding in the mangrove roots, also the upside-down jellyfish cassiopeia lying on the bottom. For a more intimate experience the centre also runs guided snorkelling tours through the mangrove.

PLACES TO VISIT

ARUBA

In addition to Arikok, Ayo Reserve with its pretty rock formations and cave petroglyphs is worth a visit.

BONAIRE

The slave huts along the coast road at the south of the island are a product of the salt industry, which began in the seventeenth century. They were supposedly used to shelter the slaves brought to the islands to extract salt from huge evaporation ponds. Today's salt works are obvious from the white salt mountains and pink ponds.

CURAÇAO

The historic capital of Willemstad with its eighteenth-century Dutch architecture is reminiscent of Amsterdam.

WHERE TO STAY

Aruba has the biggest choice of hotels, mostly located in the northwestern Hotel District, close to Eagle Beach. In Curaçao, accommodation is less focused on one area, though there is more in downtown Willemstad. Bonaire has similar options, but tends to be more expensive than the others.

FIND OUT MORE

Bonaire Tourism Corporation (Bonaire)
Tel: (599) 717 8322
Fax: (599) 717 8408
email: info@TourismBonaire.com
www.infobonaire.com
www.abc-travelguide.com

Diving:
Council of Underwater Resort Operators: www.curo.org
Bonaire Marine Park: www.bmp.org

Turtle-watching:
www.bonairenature.com/turtles

National parks: www.stinapa.org

Nature tours:
www.infobonaire.com/sightseeing.html#tourguides

Mangrove Information and Kayak Center:
www.infobonaire.com/kayaking.html

Getting around:
www.arubatourism.com/gettingaround

WESTERN CARIBBEAN

PANAMA

One of Panama's main advantages is that its developed urban areas are very close to the dense jungle, so access is not a problem. The wildlife shows a strong South American influence, with sloths, marmosets and monkeys all housed in the canopy. Among the birding highlights are the harpy eagle, five species of macaws and the quetzal, which can be found in the cloud forest of Chiriqui. Scuba divers and snorkellers will find killfish and dwarf armoured catfish, and there are black marlin off Hannibal Bank and Pinas Bay. Panama has 14 national parks, of which Darién National Park has been recognized by UNESCO for its natural diversity.

ACTIVITIES

Wildlife-watching

- The archipelago of Bocas del Toro is one of the best places to see wild three-toed sloths. On the island of Bastimentos it is possible to hire a boat and guide to navigate the shallow mangrove creek at Honda Bahía. The sloths can be seen in the low branches and sometimes come down to water level and hang out in the prop roots.
- Another treat on the island of Bastimentos is Red Frog Beach, where

it is easy to see the tiny, brilliant-red poison-dart frogs in the vegetation behind the beach.

- The forests of Bastimentos and Colón also have populations of mantled howler and capuchin monkeys.

Bird-watching

There are good opportunities in all the rainforests mentioned in this section.

- Swan Cay or Isla de los Pájaros is a favourite boat trip from Bocas del Toro town.
- The picturesque stacks and arches that lie off the island of Colón are home to red-billed tropicbirds, frigates and boobies. Late afternoon is a good time to see the birds as they arrive and settle for the evening. Check on wind and sea conditions at this exposed site before setting off.

Diving and snorkelling

- Areas of reef in the Bocas del Toro archipelago offer good diving. One favourite is off the coral cays of Cayo Zapatilla, part of the Parque Nacional Marino Isla Bastimentos.
- There is also plenty of good snorkelling around the islands. In addition to the coral patches, the mangroves are fascinating, with wonderfully coloured sponges, young reef fish and upside-down jellyfish or cassiopeia.
- Scuba diving is prohibited in the San Blas archipelago, although it's fine to snorkel.
- The one proviso about diving here is that visibility is highly variable depending on the rainfall, so get advice from the local dive operators.

Dolphin-watching

- Boat trips run out of Bocas del Toro town to visit a group of resident dolphins in Dolphin Bay, a nearby lagoon south of San Cristobal. The dolphins often come and play in the wash of the boat.

PLACES TO VISIT

- **Panama Canal.** The canal is not only a remarkable feat of civil engineering, but also significant ecologically. One of the forested islands in the 'lake', Barro Colorado, is a world-renowned centre of tropical research. The forest that forms the catchment area around the canal has some of the highest biodiversity in Central America, with populations of monkeys and bats. Capybara, iguanas and crocodiles can sometimes be seen along the banks of the canal.
- **Visitors' centres at the Miraflores and Gatún locks.** A visit to either gives an insight into the workings of the canal. The display at Miraflores includes information about the wildlife and ecology of the surrounding area.

WHERE TO STAY

Panama City has most of the major hotel chains, but travel out to Bocas del Toro, Isla Colón and Volcán Baru National Park, and you can find a good choice of eco-resorts, lodges, inns and cabins.

FIND OUT MORE

Panama Tourist Office (Panama)
Tel: 507 226 7000
Fax: 507 226 4002
www.visitpanama.com
www.panamaconsul.com
www.panamainfo.com
www.panama-guide.com

Barro Colorado STRI:
http://stri.org/english/visit_us/barro_colorado/

Diving:
www.panamadive.net
www.panamadivers.com/

Getting around:
www.panamainfo.com/en/article/visitor_information/63/

COSTA RICA – CARIBBEAN COAST

Palm trees, thick jungle, coral reef and forest characterize Costa Rica's Caribbean coast. The 18,000-ha (46,000-acre) Tortuguero National Park is a vast low-lying area covered by tropical rainforest. Tortuguero Beach is the most important nesting site of the endangered green turtle in the western hemisphere. Giant leatherback, hawksbill, and loggerhead turtles also nest here. You can kayak, boat or hike through the forest trails and spot birds such as Tinamous, brown and red-footed boobies, brown pelicans, magnificent frigatebirds and black-eared wood quails. The Cahuita National Park, south of the principal port of Limón, has one of the greatest varieties of live coral in the world.

ACTIVITIES

Turtle-watching

- The Caribbean Conservation Corporation runs guided walks on Tortuguero Beach. The seasons for turtle-watching are March for leatherbacks, March to June for greens and July for hawksbills.

Nature-spotting

- Tortuguero is a good place to see all kinds of wildlife. Behind the beach the park is a mix of rainforest and swamp. Early-morning boat trips along the lagoons and canals offer Central American spider monkeys, white-faced capuchins, mantled howlers, sloths, iguanas, basilisk lizards and common caiman, and there is even a chance of catching a glimpse of the rare Baird's tapir, neotropical river otters and jaguars.
- Monkeys can also be seen at Cahuita National Park, close to Puerto Viejo de Talamanca. The park covers an area of coastal forest, and includes a lovely sandy bay with an easy path into the forest right next to the beach. There are several troops of mantled howler monkeys and white-throated capuchins. They are used to the tourists and frequently sit relatively low down in the trees.

Bird-watching

- Tortuguero has more than 400 species of birds. The great green macaw is best spotted from December to April, when it feeds on almendro fruit. Deep in the quieter, narrow channels you may see boat-billed herons, white-fronted nun birds, many kingfisher species, black-throated trogons, slaty-tailed trogons and ant shrikes.
- From early September the annual migration from North America commences, following the Caribbean coastline, and Tortuguero is right on the flight path. Flocks of barn

swallows, eastern kingbirds and purple martins are common, while peregrine falcons appear in October.

Dolphin-watching

- Two different species are found off the coast of the Gandoca-Manzanillo National Wildlife Refuge – bottlenose and Guyanese dolphins. Some scientists think that there are hybrids, as the shape of the dorsal fin of some individuals looks halfway between the two. There are boat trips out of Manzanillo, but it's best to check locally how frequently the dolphins have been sighted before committing yourself.

Walking and hiking

- Manzanillo is the starting point for the 5.5-km (3½-mile) coastal trail to Punta Mona. It crosses a mix of swamp, forest, little coves and rocky promontories. A more demanding 9-km (5½-mile) trail leads to Gandoca and it's worth considering a guide as local conditions change at the swamp edge.
- The beach and forest walk along the bay at Cahuita National Park is flat and easy. The further away from the park entrance you go the more likely you are to lose other tourists.
- The Tortuguero Nature Trail starts at the park headquarters and ends at the beach. Other short trails start from the Jalova ranger station, at the south end of the Tortuguero National Park. They can be swampy, so again check locally before heading out.
- Generally wildlife spotting is best from the boats. There are several other opportunities for hiking and all kinds of eco-adventures such as white-water rafting, especially in the forests inland. The local tourist boards have details.

PLACES TO VISIT

- **Aviarios del Caribe.** This sloth rescue centre is 31 km (19 miles) south of Limón, and there are usually about ten sloths in the grounds. It's a good opportunity to see these charming animals close up.
- **Selvatura Park.** This cloud-forest conservation project located in Monteverde includes canopy tours, treetop walkways, butterfly and hummingbird gardens and an education centre.

WHERE TO STAY

Cities such as San José have 'super' hotels, as do the main coastal resorts. The last few years have seen a huge increase both in guesthouses and inns, and also eco lodges, B&Bs and camping facilities. Reservations are highly recommended for all forms of accommodation.

FIND OUT MORE

Costa Rica Tourist Board (Costa Rica)
Tel: 506 223 1733
Fax: 506 223 5452
www.visitcostarica.com
www.britishembassycr.com
www.infocostarica.com

Costa Rica Tourism and Travel Bureau:
www.costaricabureau.com

Caribbean Conservation Corporation:
www.cccturtle.org

Tortuguero National Park:
www.cccturtle.org/volunteer-research-programs

Selvatura National Park:
www.selvatura.com

Cahuita National Park:
www.costaricabureau.com/nationalparks/cahuita

Talamanca Dolphin Foundation:
www.dolphinlink.org

Refugio Nacional Vida Silvestre Gandoca-Manzanillo (in Spanish):
www.1-costaricalink.com/costa_rica_parks/gandoca_manzanillo_national_wildlife_refuge_esp

Getting around:
www.govisitcostarica.com/category/transportation/transportation

HONDURAS AND THE BAY ISLANDS

Honduras and the Bay Islands offer the visitor a cornucopia of Caribbean attractions – idyllic beaches with mirror-smooth turquoise waters, breathtaking landscapes and wildlife in profusion. Some of the world's best diving is available on an extension of Belize's magnificent barrier reef. For adventurous travellers, both mainland and islands have the added attraction of being less 'discovered' than many other parts of this region.

ACTIVITIES

Diving and snorkelling

- The main Bay Islands of Roatán, Barbareta, Utila and Guanaja all offer great diving, with spectacular fringing reefs and drop-offs.
- Roatán has about a dozen specialist dive resorts offering trips to local sites and longer all-day excursions. There are spectacular wall and drift dives on both north and south shores, with a cluster of sites along the island's West End and on the south coast between French Harbour and Oak Ridge.
- There are also great dives off the central-southern coast. One of the best known is Mary's Place, where vertical fault lines have cut through the reef, creating walls lined with gorgonian fans and spectacular fish.
- There is also a popular Caribbean reef shark dive on the south side of Roatán, known locally as Cara a Cara ('face to face'), because, as one diving website puts it, 'You will be literally cara a cara with the pearly white teeth of these beautiful sharks.'
- In addition to diving, Anthony's Key Resort, near Sandy Bay, offers informative dolphin encounters, including snorkelling or diving with them. The Roatán Institute for Marine Sciences is also based here and runs education programmes in marine ecology.
- The reef off Barbareta is extensive and the Elbow, which connects a series of islands called Pigeon Cays, is particularly beautiful.
- Utila has some of the most dramatic drop-offs in the Bay Islands, with walls plunging 1000 metres (3300 feet) into the deep along the north shore. Turtle Harbour Marine Reserve includes Pinnacles and Blackish Point dive sites. To the south there are numerous shallow dives and the reef flat extends to a series of coral cays. Black Hills is popular for its abundance of fish.

- Utila is also known for its seasonal pelagic visitors off the northeastern end of the island. In the right conditions it is possible to spot whale sharks, mantas and schools of bonitos. The best time of year is February to early May, though precise times are not predictable. Some of the dive operators offer whale-shark watching and swimming between dives. Utila Whale Shark Research runs educational programmes about whale sharks.
- Guanaja is surrounded by a fringing reef with coral cays on the south side. Some of the best dive sites include the Pinnacle, Jim's Silver Lode, Gorgonian Wall and Caldera del Diablo, which is famous for its spawning groupers in January and February. There are also the wrecks of the *Jado Trader, Don Enrique* and *Ruthie C.*
- There are good snorkelling opportunities on all the Bay Islands. Many of the resorts on Roatán offer boat trips to the reef flats on both shores and there are suitable shallow spots off Utila, Guanaja, Barbareta and Cayos Cochinos.

Wildlife-spotting

- Punta Sal or Parque Nacional Jeannette Kawas, as it is now known, to the west of Tornabe on the north coast of Honduras, is a good place to spot mantled howler monkeys and white-fronted capuchins. They inhabit a rocky promontory with forest and beaches. There are also seabirds on the rocks offshore.
- The peninsula of Punta Sal is accessible by boat from Miami, a small fishing village on the north coast that is part of the Jeannette Kawas National Park. There are white sand beaches, jungle and coral reef, and you can see dolphins, monkeys and birds.
- Cuero y Salado Wildlife Refuge is about 30 km (20 miles) to the west of La Ceiba on the north coast of Honduras and is an adventure to get to. You go by road as far as La Unión, then the only access is by a small train that terminates at the park office. From there you hire a canoe, preferably with a guide, and paddle silently up the river and creeks. It is possible to see white-fronted capuchins, crocodiles and mantled howler monkeys, as well numerous bird species. In total there are 35 species of mammal and 196 birds in the park.
- Each of the Bay Islands has its own species of iguana. For a guaranteed close encounter (albeit in captivity) with Utila's endangered endemic black iguana or swamper, visit the Iguana Research and Breeding Station. Other islands also have wild green iguanas.
- Behind the city of La Ceiba is the Pico Bonito National Park, a huge rainforest wilderness area with hiking trails, rivers and waterfalls. The park is home to over 325 species of birds as well as jaguars, tapirs, deer, pumas and white-faced and spider monkeys. It houses The Lodge, a luxury nature resort.

WHERE TO STAY

The bigger, more exclusive hotels are mostly in Tegucigalpa and San Pedro Sula, though mid-range and budget options can be found across the country. Most of the dive resorts in the Bay Islands are relatively luxurious places and offer weekly packages rather than nightly rates. Some of the national parks allow camping, but check in advance.

FIND OUT MORE

Honduras Tourism Institute (Honduras)
Tel: (504) 222 2124 ext 502, 503
Fax: (504) 222 2124 ext 501
e-mail: tourisminfo@iht.hn
www.letsgohonduras.com
www.honduras.com
www.hondurasemb.org

Pico Bonito Foundation in La Ceiba (for information on national park and guides): www.picobonito.com

Anthony's Key Resort coral reef research: www.anthonyskey.com

Coral Cay Conservation Project Bay Islands:
http://roatanet.com/travel/honduras-coralcay.php

For more about Utila and the Bay Islands: www.aboututila.com
www.utilawhalesharkresearch.com

Iguana Research and Breeding Station:
www.utila-iguana.de/visitors/Index.htm

For eco adventures in La Mosquitia:
www.honduras.com/moskitia

Getting around:
www.gapyear.com/honduras/getting_around

BELIZE

The tiny country of Belize is home to the world's second-largest barrier reef and three pristine atolls, providing an endless choice of dive sites. The huge diversity makes Belize one of the best dive destinations in the Caribbean. The real highlight, though, is the spectacular and perfectly circular entrance to the world-famous Blue Hole, an underwater cave fully deserving of its World Heritage Site status.

In addition, one-fifth of Belize's land is dedicated as nature reserves, and trails take you into the jungle past stunning waterfalls. Over 500 species of birds in all types of environments also make this a world-class destination for bird-watchers – just in and around Belize City 120 species can be found.

The climate can be unpleasantly sticky, though it is cooler in the mountains in the west. Heavy rainfall – almost 4 metres (160 inches) annually in the wettest areas in the south – supplies the many creeks, rivers and waterfalls.

ACTIVITIES

Diving and snorkelling

- Part of the attraction of diving in Belize is that there are nearly 400 species of fish here, 70 hard corals and huge sponges, which come in every conceivable colour and shape. Visit in August and if you are lucky you will witness the coral spawning. Triggered by the full moon and water temperature this mass spawning is one of the reef's greatest spectacles.
- The phenomenal Blue Hole has been attracting divers ever since Jacques Cousteau explored its depths in the early 1970s. Some 305 metres (1000 feet) across and 122 metres (400 feet) deep, it is found on Lighthouse Reef; to dive it from

Belize City is a day trip. To see the enormous stalactites you need to dive to a depth of 40 metres (130 feet).

- Lighthouse Reef also offers stunning drop-offs, of which perhaps the most spectacular is the Half Moon Caye Wall.
- For dives close to the shore Ambergis Caye and Caye Caulker are good bases, with the Hol Chan Marine Reserve near by.

Whale-shark diving

Gladden Spit Marine Reserve is the traditional place to see these magnificent animals. In March, April and May schools of fish such as cubera and dog snapper gather to spawn. This is a spectacle in its own right, but the huge clouds of spawn attract whale sharks that come to feast. Rising from the deep, as many as 25 sharks have been recorded at one time. The authorities encourage snorkelling rather than diving for whale-shark interaction, believing it to be less stressful to the sharks.

Manatee-watching

- Belize is a stronghold for the Antillean manatee and it is protected by law – this means you are not allowed to get in the water with them. However, lots of manatee-watching boat trips run out of Ambergis and Caye Caulker and on the half-day trips to Swallow Caye sightings are almost guaranteed.

Caves and caving

- Belize's extensive underground cave systems with their vast, cathedral-like chambers can be explored with the help of guides. A 30-minute hike uphill from just outside Xunantunich takes you to the entrance of Che Chem Ha Cave where many large, intact pieces of Mayan pottery have been left in place for visitors to see. In the Actun Tunichil Muknal cave remains of sacrificial victims have also been found.
- Subterranean river systems can be visited by canoe or kayak, but perhaps the most unusual way to explore is from the comfort of a rubber ring as you gently float down river, into the caves. Lie back and enjoy the views of the stalactite-lined ceilings as you drift by.
- The Caves Branch river system near Belmopan offers kayaking and rides into the caves on inflatable inner tubes.
- All trips into the caves should be undertaken only through guided tours.

Walking and hiking

- The Cockscomb Basin Wildlife Sanctuary is home to howler monkeys and jaguars, and has numerous trails with wooden bridges taking you over the many creeks and deep into the jungle. There are 12 self-guided trails, one of which offers the reward of a swim at the waterfall at the end.
- The Pine Ridge Forest Reserve is full of stunning waterfalls, including the spectacular Thousand Foot Falls at the heart of the reserve and numerous falls among the large boulders at Rio on Pools.
- Blue Hole National Park has many trails, picnic areas and an observation tower. You can refresh yourself after a hot hike by taking a dip in the waters of one of the blue holes found here in the rainforest.

Bird-watching

- Cockscomb Basin Wildlife Sanctuary (which was actually set up to protect the jaguar) has nearly 300 species of birds including keel-billed toucans. Here and Red Bank are good places to look out for scarlet macaws.
- A short walk through the protected area of Half Moon Caye takes you right in among a colony of red-footed boobies; a tower has been erected giving the opportunity for fantastic eye-level encounters with the birds nesting in the trees just metres away. The colony is also home to magnificent frigatebirds and if you're lucky you will see them harassing the incoming boobies and stealing the fish they have caught.
- The Crooked Tree Wildlife Sanctuary is also a must for birders. It is a wetland reserve where snowy and great egrets can be seen, as well as boat-billed herons and the jabiru stork. To get the most from your visit, go in May – when the water is at its lowest and the birds are forced to come out into the open – and take a guided tour.
- Egrets, blue herons, iguanas and turtles can be seen in the course of a canoe trip on the Sittee River.

Mammal watching

- The lush tropical forests are host to an extraordinary number of mammals – a group of animals that is conspicuous by its near absence on the majority of Caribbean islands. Coatimundi, peccary, tapir and the beautiful but rarely seen ocelot and jaguar are all found. If the animals prove elusive, however, don't overlook the many orchids, bromeliads and other epiphytes that festoon trees such as the ceiba and mahogany.

Crocodile safaris

Tours can be taken to see crocodiles on the New River. Using torches you can spot them by the reflection of the torchlight in their eyes.

Mayan ruins

The Mayan people inhabited Belize almost 4000 years ago, and 2000 years ago they began building magnificent cities and ceremonial centres. Today their legacy remains in the form of the many temples and pyramids dotted round the country. Xunantunich (Maiden of the Rock) is probably the most accessible. It is reached via the hand-pulled ferry over the Mopan River from which it is a further 1.5-km (1-mile) walk on a paved road.

- The closest Mayan ruin to Belize City is Altun Ha (Water of the Rock). The 4-kg (10-lb) jade head of the sun god Kinich Ahau was discovered here.
- The temple at Lamanai (Submerged Crocodile) is set in lush jungle and was in use up to the time that the Spanish arrived. The best way to visit is by boat up the New River, where you might even spot a crocodile.
- For Mayan enthusiasts, Caracol (the Snail) is the largest site in Belize, covering 38 sq. km (15 sq miles).

PLACES TO VISIT

- **Community Baboon Sanctuary**

It's actually the black howler monkey that this sanctuary was set up to protect. You can learn about it and the other 200 species of wildlife found here, as well as taking a guided nature tour.

WHERE TO STAY

There is a noticeable lack of the larger, glitzy hotels and much of the accommodation on offer is guesthouse style, with varying degrees of luxury. Towns in the jungle interior such as San Ignacio have some impressive plantation-style lodges, and the forest itself even has a range of stand-alone cabins. Offshore, you can find small hotels and, in all, it's a relatively cheap country.

FIND OUT MORE

Belize Tourist Board (UK)
Tel: 020 8948 0057
Fax: 020 8948 0067
www.travelbelize.org
www.embassyofbelize.com

The Belize Audubon Society:
www.belizeaudubon.org

While the name 'Audubon' suggests bird-watchers, the society is interested in all aspects of natural heritage and is dedicated to the preservation of the wildlife and natural resources of Belize.

Gladden Spit Marine Reserve:
www.friendsofnaturebelize.org/gladden_spit

Belize Eco-Tourism Association:
www.bzecotourism.org

Getting around:
www.travelbelize.org/trans

MEXICO – YUCATÁN PENINSULA

The Yucatán Peninsula is one big limestone plateau characterized by the variety of caves and cenotes that honeycomb it – ranging from dry caves such as those at Aktun Chen to amazing systems now wholly or partially submerged. Both types have incredible stalagmite and stalactite formations. The Caribbean coast of Yucatán and the offshore Isla Cozumel have some beautiful fringing reefs, and the largest area of tropical vegetation is in the 5000-sq-km (2000-sq-mile) Sian Ka'an Biosphere Reserve (the name means 'where the sky begins'), which is a designated World Heritage Site. Valladolid is a good base for the world-famous Mayan site of Chichen Itza, and not far from Tulum or Coba.

ACTIVITIES

Diving and snorkelling

- There are numerous dive facilities from Tulum to Isla Mujeres. The island of Cozumel, surrounded by coral reef, is one of the most popular diving destinations in the whole Caribbean. It has numerous shore and boat dives concentrated in the south and west of the island, and there are dozens of dive operators. Many of the highlights are wall dives, such as Santa Rosa Wall and Punta Sur Reef. Some include caves and tunnels. The currents are strong and many of the sites are drift dives.
- There are also good snorkelling spots in the shallower reefs, such as Palancar Gardens. Marine life includes barracudas, turtles, moray eels and spotted eagle rays. Elkhorn corals tend to predominate close to shore, with the largest brain and plate corals in deeper waters.

Bird-watching

- It's worth watching out for birds in the forests surrounding Mayan ruins like Coba and in the Sian Ka'an Biosphere Reserve.
- Bird-watching boat trips to an easily accessible flamingo colony at the Rio Celestún Biosphere Reserve on the west coast of Yucatán, on the Gulf of Mexico, start from the town of Celestún. April to September are the best months to visit for the flamingos, although there are many other bird species year round. The trip takes you along the coast to the estuary, then past a 'petrified forest' of trees that was killed by salt water as the sea penetrated the area.
- There is an even larger flamingo colony close to the fishing village of Río Lagartos in the north of the peninsula (although again, strictly speaking, it is part of the Gulf of Mexico). The best time to see hundreds or even thousands of flamingos is from June to August; during these months there are plenty of boat captains and naturalist guides on hand to take you to the best places and to help you spot the numerous other bird species that inhabit the mangrove estuary.

Turtle-watching

- Four species of turtle nest on the beaches of the Yucatán – green turtles, loggerheads, hawksbills and leatherbacks. Two of the preferred nesting spots are Akumal and Paamul. The best time for nesting turtles is the summer months of July and August.
- On the western beaches of Isla Mujeres there is a turtle farm where hatchlings are protected until they are tagged and released.

Whale-shark watching

Numerous whale sharks congregate each summer at Isla Holbox, in the north of the peninsula. However, as this is where the murky waters of the Gulf of Mexico meet the Caribbean, it is often hard to see them clearly.

Beaches

The Mayan Riviera, as it is known locally, has a number of beautiful sandy beaches with varying stages of development and access. Tulum is one of the best, with many kilometres of white sand. The beaches around Playa del Carmen are also easily accessible but more built up.

PLACES TO VISIT

- **Mayan ruins.** Situated right on the Caribbean coast, the walled Mayan city of Tulum has a special atmosphere. The castle and palace are perched up on a limestone cliff with a pretty, sandy beach below. Look out for iguanas basking on the ruins in the morning sunshine. To avoid the crowds, visit in early morning or late afternoon.
- **Coba** is one of the largest Mayan cities, with substantial areas still unrestored. Set in tropical forest about 50 km (30 miles) inland from Tulum, the ruins have a less formal atmosphere than some of the better known sites. They are a great place for a cycle ride, with wide paths between the various ruins, but if you don't want the exercise there are cycle taxis for hire.
- **Inland in the Yucatán**, the great ruins of Chichen Itza and Uxmal are well worth a visit. They are arguably the most famous Mexican ruins, dating back to 1000–800 AD.

- **Playa del Carmen.** A bustling resort town, with shopping, restaurants, dive operators, nightlife and a beautiful sandy beach.
- **Xel-Ha.** Meaning 'place where water is born', this is a theme park, a quick and family-friendly introduction to the habitats and creatures of the Mayan coast. You can float through clear-water mangroves on a rubber ring, snorkel among reef fish, swim with dolphins and see manatees.

WHERE TO STAY

Cancun is very touristy and has a huge range of hotels, though you can try the downtown area for a more authentic experience. There are several hotels in Chichen that would make a good base, and in Uxmal there is a pleasant lodge if you are doing the archaeological site as more than a day trip.

FIND OUT MORE

Mexican Tourist Board (UK)
Tel: 020 7488 9392
Fax: 020 7265 0704
www.visitmexico.com
www.yucatantoday.com

Caving: Association for Mexican Cave Studies: www.amcs-pubs.org

Diving:
In Cozumel – Buceo Médico Mexicano has a recompression chamber, operates a 24-hour emergency service and keeps a list of affiliated dive operators.
Tel: 52 987 2 1430
www.hiddenworlds.com.mx

Mayan Riviera: www.mayanriviera.com

Getting around:
www.mexperience.com/guide/essentials/getting_around

OUTER CARIBBEAN

BAHAMAS

Best known for its idyllic beaches and one of the most sophisticated tourist set-ups in the Caribbean, the Bahamas are also a mecca for divers. These islands lay claim to the world's third-longest reef – some 225 km (140 miles) long, its lies anything from a few hundred metres (yards) to 3 km (2 miles) off the shore of the largest island, Andros. Many of the islands are pitted by blue holes, the circular sinkholes that are a window on the submarine cave systems.

ACTIVITIES

Swimming with dolphins

- A wide variety of organized trips allows you to interact with the Atlantic spotted dolphins and bottle-nose dolphins that are resident in the warm, clear Bahamas waters. Scuba divers may even find that adolescent dolphins from the larger pods 'socialize' with them.
- For non-divers there are snorkelling options and even trips to protected shallow lagoons, where you can wade out to meet the dolphins, which are so used to humans that you may get close enough to touch them.
- Some operators also offer longer live-aboard excursions where you stay on a boat for up to a week, giving you the opportunity for plenty of magical dolphin encounters.

Canoeing

- With kilometre upon kilometre of coastline to pick from you can explore the wetlands and mangroves as you glide silently along in a stable sea kayak, accompanied by an expert guide. Around Cat Island you can see rays and sharks in the waters while looking out for egrets and white-crowned pigeons.
- Should you get the kayaking bug there are week-long trips available on two-person kayaks. At the end of each day you get to camp out on a beautiful beach either in a tent or under the stars.

Walking and hiking

- Few people come to the Bahamas simply to hike and there is not a well-developed trail system as yet, though the remote, pine-covered Abaco National Park on the island of Great Abaco is good when you've had enough of the sea. Look out for the Bahamas parrot (see Bird-watching, p.220). Beware, poisonwood is also common and it can give a rash to the unwary hiker brushing against it.
- Reached by a long, arduous and pot-holed road through the park is the Hole-in-the-Wall lighthouse. For those determined enough to get there the classic red-and-white-striped structure presides over some fantastic views of this wild and remote headland.
- The Lucayan National Park on Grand Bahama has a variety of short trails through mangroves.

Diving and snorkelling

- With over 700 islands scattered over 160,000 sq. km (100,000 square miles) of sea, there is no shortage of diving in the Bahamas. Thousands of sites offer a huge range of experiences from shallow reefs to drop-offs, tunnels and crevices to coral heads spread out over a vast arena of white sand.
- The islands' many blue holes may descend as deep as 180 metres (600 feet). Lucayan National Park has one of the world's largest explored underwater cave systems but it should be dived only by experienced divers accompanied by qualified guides.
- The cays and islands of the Exuma Cays Land and Sea Park provide excellent dives. Staniel Cay is home to the Thunderball grotto, a beautiful setting that has appeared in its 007 namesake and other films.
- Pelican Cays Land and Sea Park protects over 770 ha (2000 acres) of sea and small cays off the island of Great Abaco. It is good for both snorkellers and divers, with shallow reefs that help explain why these reefs are often referred to as underwater gardens. There are land-based dive operations and live-aboard dive boats to cater for all tastes.
- One of the world's only two stromatolite or 'living fossil' reefs is found off the Atlantic coast of Stocking Island in the Exumas. It is said to date back over 3.5 million years and is an odd and unique dive.
- Shallow waters combined with the many wrecks found here make the Bahamas excellent for wreck divers and most of the islands have wrecks of some kind to explore, from Spanish galleons complete with cannons to old

Bond-movie sets, including the famous Vulcan Bomber site from *Thunderball*. Good wreck dives include the *Theo*, a 73-metre (240-foot) freighter off the shores of Grand Bahama, and Wreck City, on New Providence's southwest corner, which offers a dozen deliberately sunk wrecks in close proximity, including those left behind by Hollywood film-makers.

- Night dives show a different side to the wrecks – on some, sleeping turtles can be found.

Diving with sharks

- You can go on organized shark dives where Caribbean reef sharks are attracted in large numbers when they are fed by the dive operator. The sharks can be quite boisterous and will bump divers. Although not quite the 'natural' encounter you might wish for, it is still exhilarating to be surrounded by these top predators and you may end up feeling grateful that you are not alone during the experience.

Bird-watching

The Bahamas is a bird-watcher's paradise, with over 200 species and the open nature of the countryside offering a chance to see birdlife on every island. If you want to tick off one of the few endemics, look out for Bahama parrots in the pine forests of Abaco National Park.

- The Bahamian hummingbird is more widespread, found all over the Bahamas, as is the stunning ruby-throated hummingbird.
- Bird-watching trips can be taken with guides from the Bahamas National Trust, who also organize a week-long 'Birding in Paradise'.
- Great Inagua, the southernmost island in the Bahamas chain, is well off the beaten path but worth a special trip. Here, in Inagua National Park, is the world's largest breeding colony of Caribbean flamingos. A hostile environment for humans, it is nevertheless a perfect place for the birds. Large, salty pools teem with brine shrimp and larval brine flies on which they feed.

Flowers

Despite the lack of tropical rainforests, there is no shortage of colourful vegetation, with every month having something in flower. The blue mahoe is an endemic hibiscus with yellow to red flowers; there are also flowering cactus. Many of the plants found throughout the region are used for bush medicine, and bush-medicine tours are available.

PLACES TO VISIT

- **Mount Alvernia Hermitage.** On Cat Island is the former home and hermitage of the architect, missionary and Franciscan monk Father Jerome Hawes. This miniature medieval-looking monastery, built in the 1940s, is on the Bahamas' highest point and as well as being a curious building it offers great views.
- **The Retreat.** This 4-ha (11-acre) garden on New Providence is home to the Bahamas National Trust and one of the world's largest collection of palms. Other specimens include the local mahogany and many native orchids.
- **Botanical Gardens.** Also on New Providence, the Botanical Gardens have many tropical species as well as ponds, waterfalls and grottoes. An unusual feature is the creation of a Lucayan village with traditional thatched buildings.
- **Fort Charlotte.** Finished in 1790, the largest fort in the Bahamas guards Nassau harbour. Tours are available to visit the dungeons, tunnels and bombproof rooms.

WHERE TO STAY

The Bahamas probably have one of the most comprehensive accommodation choices in the Caribbean, with luxury, all-inclusive resorts as well as large hotels. Privately rented accommodation is also plentiful. There aren't too many camp sites, but there are several dive resorts and places to stay near the national parks.

FIND OUT MORE

Bahamas Tourist Board (UK)
Tel: 020 7355 0800
Fax: 020 7491 9459
www.bahamas.co.uk

Bahamas Ministry of Tourism (Bahamas)
Tel: 242 302 2000
Fax: 242 302 2098
Toll Free: 1-800-Bahamas
E-mail: tourism@bahamas.com
www.bahamas.com

Bahamas National Trust:
www.thebahamasnationaltrust.org

Diving:
Stuart Cove: www.stuartcove.com
UNEXSO: www.unexso.com

Getting around:
www.bahamas.com/bahamas/about/general.aspx?sectionid=26021

TURKS & CAICOS ISLANDS

These two island groups are in the North Atlantic Ocean, southeast of the Bahamas. Grand Turk has small local settlements and historic ruins – fewer than 5000 people live on this 19-sq.-km (7½-sq-mile) island. The islands are arranged around two large limestone plateaus, the Turks Bank, with deep offshore waters that host humpback whales, spotted eagle rays, manta rays and turtles. Over 190 species of bird can be found, 52 of which are known to breed locally. Ruddy turnstones, least sandpipers, greater yellowlegs and lesser yellowlegs can often be found feeding together in the winter months.

Find out more:
www.turksandcaicostourism.com

FLORIDA KEYS

From Key Largo to Key West, some 1700 islands stretch out from the southern tip of Florida, essentially exposed portions of a coral reef. Tidal creeks and mangrove forests are easily explored, and kayakers can spot jewfish, loggerhead turtles and lemon sharks. There is an abundance of nature trials, lined by tropical plants and hardwood trees. On the peripheries of the woods, birders can spot mangrove cuckoos and white-crowned pigeons. Key Largo houses John Pennekamp Coral Reef State Park, the first undersea park in the USA, which, together with the Florida Keys National Marine Sanctuary, covers 178 nautical miles of reef and mangrove swamp.

Find out more: www.fla-keys.com

Published in the United States in 2007 by Yale University Press.
Published in the United Kingdom in 2006 by BBC Books, an imprint of Ebury Publishing.
Ebury Publishing is a division of the Random House Group Limited.

Commissioning Editor: Shirley Patton
Project Editor: Eleanor Maxfield
Copy Editor: Caroline Taggart
Designer: Bobby Birchall
Production Controller: Peter Hunt

Set in Officina Sans and Futura
Printed and bound in China by C&C Offset Printing Co., Ltd.
Colour separations by Dot Gradations Ltd., UK

Library of Congress Control Number: 2007920724
ISBN 978-0-300-12549-8 (pbk. : alk. paper)

A catalogue record for this book is available from the British Library.

The paper in this book meets the guidelines for permanence and durability of the Committee on Production Guidelines for Book Longevity of the Council on Library Resources.

10 9 8 7 6 5 4 3 2 1

PICTURE CREDITS

Barrie Briton: TR 48, 80, 115, **Martyn Colbeck:** TL 48, **Juan Carlos:** 34 **Eyeubiquitous/Hutchison:** Paul Seheult p8–9, Isabella Tree 91, **FLPA:** Reinard Dirschell 59, Michael Gore 180–1, Chris Mattison 162, Minden Pictures: Gerry Ellis 78, Tim Fitzharris 160, Konrad Wothe 54, Norbert Wu 42, **Getty Images:** Steve Winter/National Geographic 26–7, **Chris Mattison:** 24, **Naturepl.com**: Julia Bayne 112–113, Nigel Bean 11, 20, Bristol City Museum 39, Dan Burton 175, John Concalosi 82, Mark Carwardine 58, Brandon Cole 32, 68, 166, Jurgen Freund 169, Laurent Geslin 147, Nick Gordon 142, Gavin Hellier 101, Paul Hobson 144–5, Thomas Lazar 93, Neil Lucas 35, 138, David Kjaer 161, Barry Mansell 179, Luiz Claudio Marigo 110, 153, Rolf Nussbaumer 61, Pete Oxford 116, 122, 126, 148, Doug Perrine 18, 28, 30, 47, 73, 96, 104, 135, 172, 185, 186, 188, 193, Constantinos Petrinos 66, Michael Pitts 12, 74–5, 136, 163, 177, Michael Potts 15, 31R, 38, 41, Premaphotos 121, Todd Pusser 31L, 183, Jean E Roche 85, Gabriel Rojo 106, Jeff Rotman 50, 189, Francois Savigny 124–125, Lynn M Stone 108, Kim Taylor 120, Nick Upton 129, Tom Vezo 14, 70, 76, 196, Dave Watts 99, 156, David Welling 89, Doug Wechsler 45, 60, Staffan Widstrand 159, 197, Mike Wilkes 53, Solvin Zankl 17, 182, 184,
NHPA: James Carmichael Jr 1466, Daniel Heuclin 71, Trevor McDonald 49, **Mike Pitts:** 2, 13, 62, 64, 72, 140-141, 154, 167, 198 **Travel Ink:** Andrew Brown 100, Abbie Enock 94 Karen Bass: 118–19, 133

ACKNOWLEDGEMENTS

Like any book accompanying a television series, there is an entire production team behind the project, so the authors would like to thank especially: Assistant Producer Jo Ruxton and Series Researchers Stephen Lyle and Susan Gibson for access to their research; Production Co-ordinators Vicky Knight, Kim Heron and Ali Serle for their help in keeping track of production stills; Dave Williams, Juan Carlos Ocana Martinez for help with the gazetteer; Kate Pink for locating such stunning photographs; Ci Taggart for recovering meaningful sentences of the English language; Bobby Birchall for an eyecatching design; and Eleanor Maxfield for keeping the whole thing on the road and heading in the right direction.